Napier

With Best Wishes

Peter — January 1981.

THE FUTURE OF OIL

The Future of Oil

A Simulation Study of the Inter-relationships of Resources, Reserves and Use, 1980-2080

Peter R Odell
and Kenneth E Rosing

Erasmus University, Rotterdam

Kogan Page, London/Nichols Publishing Company, New York

First published in Great Britain in 1980 by
Kogan Page Limited
120 Pentonville Road, London N1 9JN

British Library Cataloguing in Publication Data
Odell, Peter Randon
The future of oil
1. Petroleum industry and trade
I. Title II. Rosing, Kenneth
339.4'8'6655 HD9560.5
ISBN 0-85038-344-7

Published in the United States of America in 1980
by Nichols Publishing Company, Post Office Box 96,
New York NY 10024

ISBN 0-89397-097-2
Library of Congress Catalog Card Number 80-24393

Printed in Great Britain by
The Anchor Press Limited
Tiptree, Essex

To the memory of Hester Rosing, 1946-79

Contents

Preface and Acknowledgments

Over the last decade the view on the future availability of oil resources and the speed with which these can be made available has changed; as has the ownership of the world's reserves and thus the likelihood of their being produced in the short to medium term; and there have also been changes in the rate at which oil is being used. The consequences of the relationships between these fundamental aspects of the future of oil are therefore also subject to radical change. This book and the research on which it is based represent a preliminary attempt to describe the way in which resources, additions to reserves and oil use inter-relate at the world level and the results for the future of oil if a wide range of values is used for the variables involved.

All the recent interpretations of the future of oil deal with a narrow range of values for these variables and thus the results stand a high chance of being invalid. For example, there are no studies on the future of the oil industry which have used a growth rate in oil use of less than three per cent per annum. Yet, since 1973, the world-wide annual average use of oil has increased by no more than 1.5 per cent. Moreover, in view of the following factors the likelihood that even this relatively low rate of increase in the use of oil will be maintained has been much reduced. First, there are oil supply difficulties in the short term (the consequence of politico-economic considerations); second, it is necessary to take account of the impact of a price for oil which is now an order of magnitude greater, in real terms, than in 1970; and, third, there will be the result of the serious efforts which are now being made by governments in many parts of the world to reduce the amount of oil being used (both through conservation and substitution measures).

On the potential supply side, recent global studies on the

future of oil have usually assumed a remaining availability of about 1600 x 10^9 barrels (additional, that is, to the 400 x 10^9 barrels of oil already used). These presentations have suggested that a resource base 50 per cent greater provides an extreme value for the variable. In fact, these views of the future potential are conservative even in respect of conventional oil, particularly with oil now valued at up to $40 per barrel, which means that more investment in exploration and exploitation is worthwhile. They also fail to take into account the potential availability of much larger quantities of oil which are now becoming recoverable from different types of habitats, *viz* the world's resources of very heavy conventional oil, such as the oil of the Orinoco Oil Belt in Venezuela and the so-called unconventional oil of the tar sands and the oil shales. Such resources of oil will be produced in increasing quantities if prices remain at present levels or go higher.

An understanding of alternative and most likely futures for oil is important in many ways. First, because questions of its supply and price are central to the prospects for the overall development of the world's economic systems and thus for the welfare of most of the world's inhabitants. Second, because the future of oil is increasingly influential in determining major political issues at the international level, including the question of peace and war. And, third, because acceptable alternative energy sources will be very expensive to develop, compared with the costs involved in oil exploration, exploitation and use, so that a premature and/or too rapid attempt to introduce such alternatives could well serve to create difficulties in the economic and social systems. This could arise if too many of the world's limited financial, managerial, technological and other resources had to be devoted to the exploitation of alternative energies so that too few resources were then available for investment in a wider range of necessary and/or desirable objectives in a much broader economic and social framework. The conclusion of this study is that these problems, which arise essentially from a belief in the inevitability of a scarcity of oil, *can* be avoided.

We have, over many years, attempted to familiarize ourselves with the issues involved concerning the future of oil through the wide-ranging literature on the subject. We have also participated in many international energy meetings with colleagues from many parts of the world across a wide variety of disciplines. The future of oil is an international and an interdisciplinary subject and we thus owe a great debt of gratitude to the many people who have contributed to our understanding. Our thanks go to all those who have – either knowingly or unknowingly – participated in the evolution of this study. We hope that our efforts may stimulate other work on the subject, particularly in respect of the costs involved in achieving various levels of oil resource discovery and the varying speeds of its development, as well as in respect of the price elasticity of the demand for oil. It is only as a result of such work that the world's statesmen, equipped with all the relevant facts and interpretations, will be able to take critical decisions on the future of oil.

Most specifically we would like to thank a number of people who have collaborated with us in producing this book. First, José Ligthelm who, as a student-assistant in Economic Geography in the Economic Faculty at Erasmus University, was responsible for processing much of the computer output on which the conclusion of this study is so dependent. Second, to our colleague, Ing. E.W. Gerritsen, for his help in solving systems difficulties in the computer work. Third, to Chris Moore, cartographer in the Economic Geography Institute and to Emile van Dijk, his assistant, for the preparation of the illustrations and to Rob Leusink who has been responsible for their photographic preparation and presentation for publication. Fourth, to Mrs E. van Reijn for her excellent preparation of the typescript at a time when she had many other tasks to fulfil as secretary to the Economic Geography Institute. And, finally, to Ethel de Keyser and Carol Steiger who had editorial responsibility for the book at Kogan Page Ltd. Their enthusiasm in working so effectively on the manuscript has ensured its publication not only in a more readable form, but also in double-quick time.

Peter Odell once again wishes to thank his family, and especially his wife, for their support and forbearance over the lengthy period during which his attention has been devoted to questions concerning the future of oil. The book is dedicated to the memory of Hester Rosing, Ken Rosing's wife. She had worked on earlier oil studies with the authors and her untimely death in February 1979 was an incalculable loss to Ken Rosing and deprived us both of a most valued colleague. Had she lived she would have been much involved in the evolution of this book and, we believe, would have approved its objectives.

Peter R. Odell
Kenneth E. Rosing

Rotterdam, August 1980

List of Illustrations

List of Illustrations

List of Tables

Chapter 1
The Basic Complexity

A. Introduction

Throughout the Western world energy policies are now based on an assumption which has been elevated to the status of a self-evident truth, *viz* that the future potential demand for oil will exceed the supply that can be made available by the world oil industry. Over the next few years there is a high probability that this will be the case, given the control that the member countries of OPEC effectively exercise over the supply by their domination of the international oil market.[1] Such politico-economic considerations can be set aside in respect of the medium- to longer-term. For the period beyond the mid-1980s predictions of a relative scarcity of oil are related to another set of factors entirely.

These are: first, a general acceptance of the hypothesis that the world's oil resource base (that is, the total amount of ultimately producible reserves of oil in the world's crust) is too small to enable production to expand for more than another few years; and, second, the belief that the annual quantity of oil which can be proved (that is, the oil which can be shown to exist and to be recoverable with existing technology and at present costs and prices) will act as a constraint on the annual rate of production. These two considerations are not, of course, mutually exclusive. Indeed, they are generally presented as complementary aspects of the oil scarcity syndrome which, implicitly if not explicitly, is also presented as a function of a third major consideration – namely, the expected continuation of high rates of increase in the use of oil.

1. See P.R.Odell, *Oil and World Power*, Penguin Books, Harmondsworth, 5th Edition, 1979, for a description of the international oil system in general and, in Chapter 9, an analysis of the evolution of OPEC control, in particular.

Given that resource, production and demand considerations all play a role in determining the future of oil, it is difficult to see why the complex future of the commodity is viewed as one which has a self-evident solution about which there is little scope for argument and even less room for divergent opinions. The complexity which necessarily arises from the inter-relationships of the variables suggests that any 'self-evident solution' ought to be treated with a healthy degree of scepticism. The need for such a sceptical reaction is strengthened if one notes the multi-faceted nature of each of the three basic variables. Each has physical, technical and economic (as well as political) aspects to its evolution and there is thus plenty of scope for differing opinions on the importance which should be attached to the various aspects and the way in which they relate to each other.

B. The Resource Base Consideration

The complexity of the resource base consideration can be demonstrated from the history of the industry in the period between 1950 and the early 1970s. This is true even though the period was one of relatively straightforward and continuing expansion of the world-wide industry, mainly under the aegis of the international oil companies, whose organization and motivations ensured the diffusion of knowledge and of advancing technology to most parts of the world.

Until the early 1970s these companies expressed little or no concern for the ultimate size of the world's oil resource base. They simply considered it to be so large as to make it irrelevant to questions of the development of the industry.[1] Nevertheless, a number of estimates were made over those years.

Twenty-three such estimates, made in the years between 1941 and 1975, are shown in Figure 1.1. This diagram shows, first, the steadily changing views on the question as estimates increased from under 500 x 10^9 barrels to over 3000 x 10^9 barrels. Second, it shows how the changing size of the estimates can be closely correlated with the time at which the different estimates were made. A correlation between the size and the date of the resource base estimates may, at first sight, appear to be of little interest and of even less significance. Time, however,

1. This point will be demonstrated in Chapter 3 when we look at the industry's pre-1973 view on the future of oil.

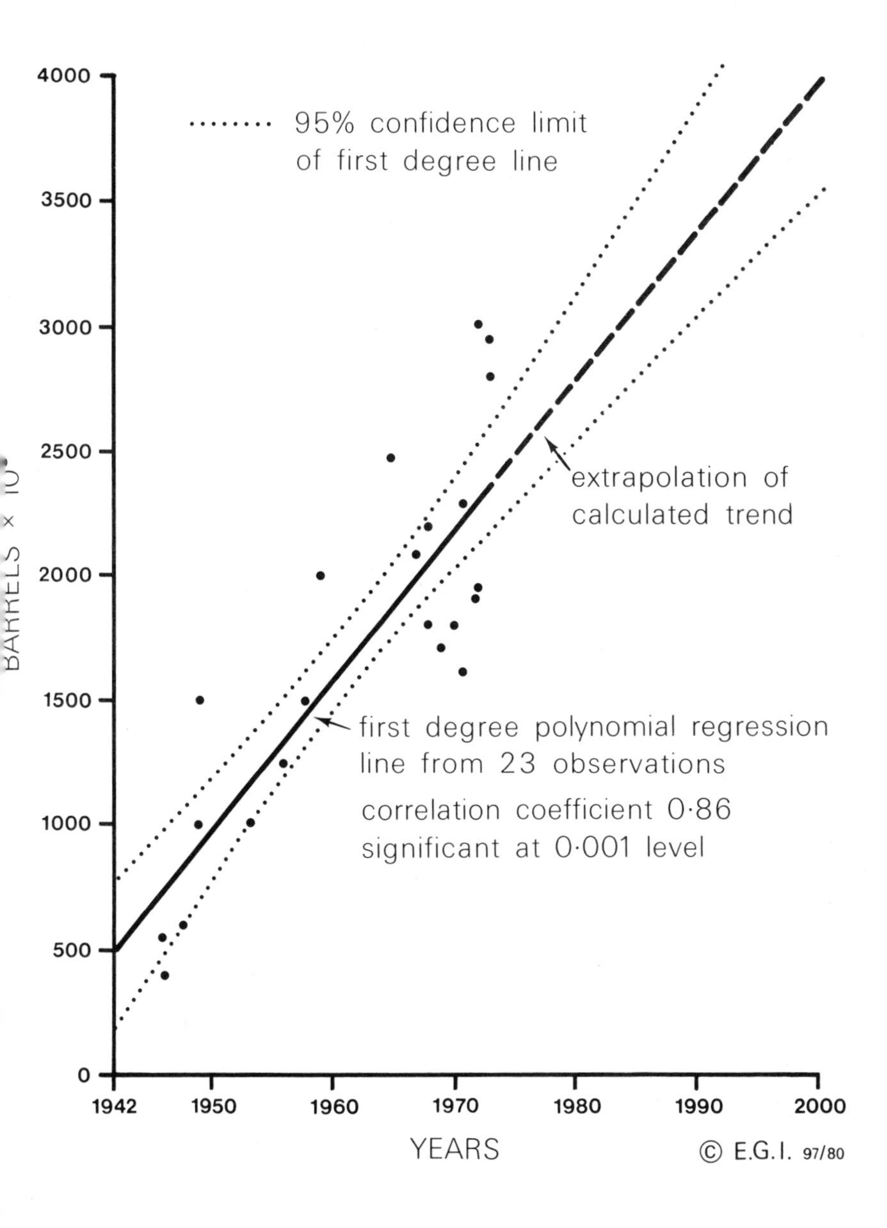

Figure 1.1 *Estimates of the World's Ultimate Oil Resources, 1942-75*

is simply the proxy variable which represents increasing knowledge and improving technology. It is on these bases that man's ability to understand and to exploit the world's hydrocarbon resources has gradually been developed and expanded. As there is, as yet, nothing to indicate that man has perfected either his knowledge or his technology, it would be unscientific to assume that the most recent forecasts of the world's ultimately recoverable oil are the last word on the subject. On the contrary, both knowledge and technology concerning the occurrence and development of oil resources continue to progress. This can be seen, for example, in the contemporary reappraisal of more extensive and deeper geological habitats of oil and in the appraisal, for the first time, following new technological developments, of oil deposits beneath the extensive continental slopes lying in water depths of 200 to 2000 metres. There thus remain general geological and technical factors which will keep the estimates of ultimately recoverable oil moving along. In addition, there are continuing opportunities for specific reappraisals of volumes of producible oil in respect of recently discovered, and as yet little known, major oil provinces – of which the North Sea and south-east Mexico are two important contemporary examples. The next 20 years appear to offer a high probability of the continuing evolution of ideas on the size of the world's ultimate oil resource base. There is, moreover, one important additional factor which will serve to encourage the process. This is the, as yet, little appreciated impact of recent increases in the price of oil on the calculation of the oil in place that it is economic to produce.

Figure 1.2 shows the formidable and continuing decline in the price of oil (in constant 1974 $ terms) over the period 1950-70. By 1970 oil was over 60 per cent cheaper than it had been 20 years previously. Yet it was despite this powerful economic disincentive to seek new oil, or even to think seriously about future, less accessible and inherently more difficult habitats and their exploitation, that the increase in estimates of the size of the world's oil resource base from 1000×10^9 barrels to at least 2000×10^9 barrels occurred. Figure 1.2 also shows the even more formidable oil price developments since 1970. In the last decade the real price of oil has increased by more than an order of magnitude in response to the revolution in the world of oil power.[1] In the middle of 1980 the price of oil is still

1. P.R.Odell, *Oil and World Power*, op. cit., Chapter 9.

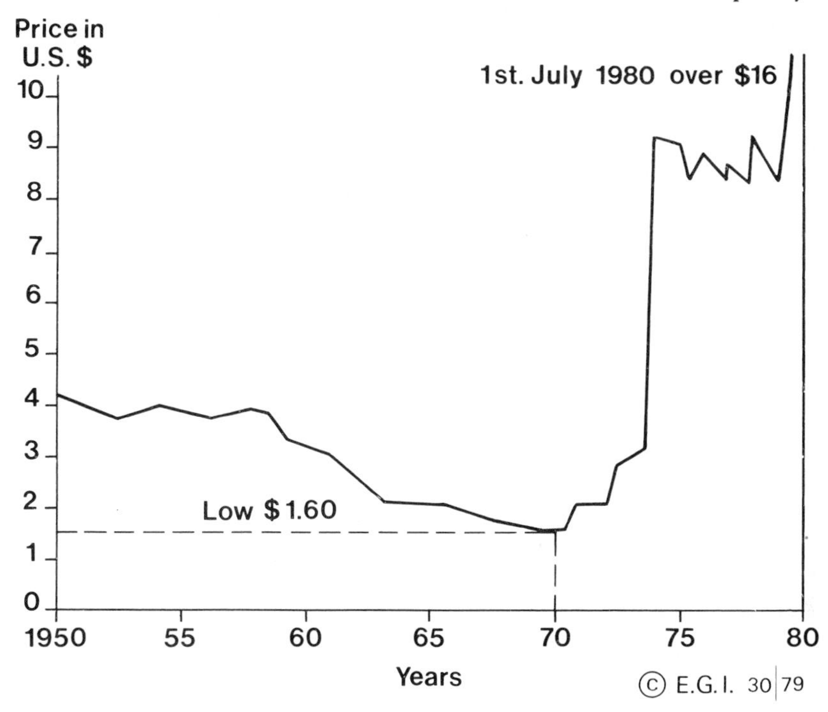

Figure 1.2 *The Price of Oil, 1950-80, based on the f.o.b. price of Saudi Arabian light crude oil expressed in 1974 $ terms*

Note, moreover, that Saudi Arabian crude oil is currently priced below comparable crudes from other countries – both OPEC and non-OPEC. A weighted average July 1980 price for all crudes of this quality would give a 1974 $ price of almost $20.

firmly on the way up to price levels which create the possibility of profitable investment in a wide range of enhanced recovery projects, and in the exploitation of the least accessible resources. This is reflected in the present situation in which most of the world's nations are anxiously, or even desperately, seeking to test for the potential availability of indigenous oil resources in order to open up the possibility of eliminating the high cost of their oil imports, and the penalties this inflicts on their prospects for economic development.

In the changed economic circumstances a significant upward re-evaluation of the world's conventional oil resource base seems to be mainly a question of the time which is necessary to organize and undertake the research and to make the re-calculations. A straight-line extrapolation through to the year

2000 of the regression shown in Figure 1.1 thus seems, under the conditions noted, as being most likely to represent a conservative statement of the future development of estimates of the ultimate world oil resource base.

To these estimated quantities of conventional oil must also be added the potential for oil resources from unconventional habitats. These are geographically extensive and include the tar sands of the province of Alberta in Canada, the heavy oil belt of the Orinoco region of Venezuela and the oil shales of the United States, Brazil, India and Malagasy etc. High production costs and low oil prices have hitherto inhibited the inclusion of unconventional oil resources in the world oil resource figures. Now, developing production technologies, coupled with the very much higher market value of oil, convert large quantities of unconventional oil into an effective resource. The volume of this addition to the ultimate oil resource base is a minimum of 2000 x 10^9 barrels[1] and a maximum of unknown dimensions, given that, to date, there has been no formal search for unconventional oil and no systematic evaluation of its occurrence on a world-wide basis.

The rapid, and more or less continuing, increase in size of the estimates of the world's ultimate oil resource base over the past 40 years demonstrates the inappropriate nature of an approach to the future of oil which employs just one or two values for this important variable. The even more recent traumatic political and economic changes in the structure of the oil sector of the world's economy justify even greater hesitancy in working with a single measure of the world's ultimately recoverable oil resources. Yet, in spite of such overwhelming reasons for extreme care in dealing with the value of the variable, there has been a marked tendency in recent forecasts to present energy scenarios with but a single future oil availability component – normally a figure of 2000 x 10^9 barrels of oil inclusive of the oil already used (about 400 x 10^9 barrels).[2] The use of this extraordinarily low figure, which also has a low probability attached to it for calculating future oil supplies to the world energy market, appears to have emerged from exclusively international oil company-inspired views of the future of conventional

1. This is a figure which is generally agreed as representing the minimum potential. See Executive Summaries of World Energy Conference Conservation Commission Report, *World Energy Resources 1985-2020*, IPC Science and Technology Press, Guildford, pp.15-16 and pp.46-7.
2. Examples of approaches to the future of oil incorporating an oil component of this size are given in Chapter 4.

oil.[1] Certainly most of the international oil companies now appear to work – in public at least – with a 2000 x 10^9 barrels oil resource base as a rule of thumb approach to the question. As will be shown later (in Chapter 3), this is a very recent specification by the companies. Moreover, as many spokesmen for the companies have themselves made clear, the figure is one which reflects their expectations as to their ability and/or willingness to explore for and to exploit oil resources in the future politico-economic conditions which they anticipate in various parts of the potentially petroliferous world. Their expectations in this respect are not at all optimistic, in view of the deterioration of opportunities and conditions for pursuing oil exploration and production activities over the last 10 years in many parts of the world.[2]

For example, the companies have largely discounted the Latin American potential for hydrocarbons. This appears to be the result of a more than 50-year history of general hostility to the companies by most countries of the continent. Over this period the companies have generally been viewed as the prime agents of US economic imperialism and have been nationalized or otherwise severely constrained in their activities, particularly in respect of their exploration and production work. As a result of this background the international oil industry's evaluation of Latin America's ultimate resource base has become unenthusiastic and outdated. Its inclusion in their world estimates of ultimate resources at a level of 150-200 x 10^9 barrels reflects this situation. This has to be compared with alternative views of others such as those of Dr B.Grossling (formerly of the United States Geological Survey [USGS] and now with the Inter-American Development Bank), who gives a

1. Note, for example, the number of oil company estimates of ultimate oil availability at about this level in the delphi-type survey on the subject undertaken by the Institut Français du Pétrole for the 1978 World Energy Conference, *op. cit.*, pp.6-11 and 24, and the official presentation by the Mobil Oil Corporation (at the 1979 World Petroleum Congress) of an oil reserves base of this size. See *Petroleum Economist*, Vol.XLVI, No.12, December 1979, pp.501-2.
2. See, for example, the comments by R.A.Sickler (Royal Dutch/Shell Exploration and Production Division) in *Methods and Models for Assessing Energy Resources* (Ed. M.Grenon), Pergamon Press, Oxford, 1979, pp.137-140. In response to questions and discussion on the background to the presentation in his paper of the idea of a 2000 x 10^9 barrels ultimate oil resource figure, Sickler eventually agreed that he was *not* presenting an estimate of ultimately recoverable resources. These, he agreed, 'have been constrained logistically in light of the climate (for exploration and development) available now and the scenario (for the increase in reserves) seen now up to about the year 2000'.

range for Latin America's ultimate oil resource of 490 to 1225 x 10^9 barrels, and the estimates of the Soviet Ministry of Geology of 620 x 10^9 barrels.[1] Recent oil discoveries, of major significance by world standards, in the southern part of Mexico, from which country the international oil companies were excluded over 40 years ago when the industry was nationalized in 1938, indicate that Mexico alone may well have ultimate reserves in excess of the oil companies' current expectations for the whole of Latin America.

Elsewhere in the Third World, the alternative estimates of ultimate reserves are several times higher than those currently included by the oil companies in their world-wide figures for the regions concerned. These contrasts are detailed in Table 1.1. The differences emerge from the same factors as in the case of Latin America but accentuated by the even greater paucity of oil industry efforts to find oil in these regions (see below p.43 and Figure 1.6).

Table 1.1 *Estimates of the ultimate oil resources of the Third World (in bbls x 10^9)*

	Oil Industry Views (a)	Grossling (USGS) (b)	Min. of Geology USSR (c)
Latin America	150–230	490–1225	620
Africa	120–170	470–1200	730
South /S.E. Asia	55– 80	130– 325	660
Totals	325–480	1090–2750	2010

Sources:

(a) Based on figures in R.Nehring, *Giant Oil fields and World Oil Resources*, Rand, Santa Monica, June 1978. Table on p.88 adjusted to give comparable geographical coverage with (b) and (c) above. Nehring also states (p.88) that his figures for these regions are 'roughly similar' to those published elsewhere by oil industry observers.

(b) B.F.Grossling (USGS), 'In Search of a Probabilistic Model of Petroleum Resources Assessment' in *Energy Resources*, M.Grenon, Ed., IIASA, 1976.

(c) Visotsky, V.I. *et al.*, Ministry of Geology, Moscow, Petroleum Potential of the Sedimentary Basins in the Developing Countries', *ibid.*

Thus, the figure for the world's ultimately recoverable resources of oil of 2000 x 10^9 barrels, though used almost exclusively in contemporary Western world presentations on the future of oil, is by no means a generally accepted one. 2000 x

1. See the notes to Table 1.1 for the sources for these estimates.

10^9 barrels must, indeed, be interpreted as representing the most conservative view of the availability of oil. It provides the lowest figure in a range of estimates on the world's ultimate oil resources. In a recent article Academician Styrikovich of the USSR indicated the possibility that ultimate world oil resources will be about 11,000 x 10^9 barrels.[1] This figure appears to give an upper limit to the current range of estimates though it should be noted that Styrikovich himself, later in his article, referred to this estimate as a 'cautious' one.[2] Between these extremes there have been many other estimates. In Figure 1.1, for example, there are estimates made in the early 1970s for a resource base of about 3000 x 10^9 barrels.[3] More recently, in the delphi-type survey of oil resource expectations undertaken by the Institut Français du Pétrole in 1978, there are four estimates of 4000 x 10^9 barrels or more.[4] Elsewhere an International Institute for Applied Systems Analysis (IIASA) 'workshop' on future oil and gas supply, attended by experts from many parts of the world, concluded that 'resources (of conventional oil) to be reckoned with are no longer around 3000 x 10^9 barrels, as previously estimated, but rather 4000, 5000 or even 6000 x 10^9 barrels, depending on the price which consumers can afford to pay'.[5]

In addition to the resources of so-called conventional oil there are, as indicated earlier in the chapter, also those resources of oil which are potentially available from other habitats. Production limitations, arising from technical, environmental, and economic considerations, constitute effective short- to medium-term constraints on the development and production of these non-conventional oil resources. However, when the issue under consideration is the long-term outlook for the supply of oil then it becomes quite appropriate to assume that these constraints will be diminished or even overcome. This will happen over a period of the next 20 to 30 years, assuming that

1. M.A.Styrikovich, 'The Long Range Energy Perspective', *Natural Resources Forum*, Vol.I, No.3, April 1977, pp.252-3.
2. *ibid.* p.254. He refers to an alternative Soviet view which puts recoverable resources of conventional oil at up to 15,000 x 10^9 barrels.
3. These are the estimates of the National Petroleum Council in Washington, 1972. A.R.Linden and J.D.Parent of the US Gas Association in 1973, and L.G.Weekes in a report prepared for the United Nations in 1973.
4. Institut Français du Pétrole, *op. cit.*, p. 24.
5. R.Seidl, 'Oil: The Picture is Changing', *Options*, International Institute for Applied Systems Analysis, Laxenburg, Winter 1977, p. 4.

there is a continuing demand for oil in increasing quantities during this time. This means that the potential ultimately recoverable availability of non-conventional oil can be treated as a valid input in a study of long-term future oil supply/demand relationships.

High estimates of the potential for non-conventional oil still remain speculative, as insufficient geological and technological work has been done to substantiate its existence and recoverability. There is, however, firm knowledge of the existence of 2000 to 5000 x 10^9 barrels of ultimately recoverable unconventional oil. Successes to date in recovering non-conventional oil in Canada, Venezuela[1] and the USSR indicate a high probability that a volume of resources lying within this range will become economic to produce over the next half century.

In the light of the alternative, higher estimates of the ultimate availability of conventional oil coupled with a reasonable expectation for the rapid evolution of an effective resource base of additional oil from unconventional habitats, there would seem to be quite adequate justification to simulate the long-term future of oil in a model in which the value of the resource base component rises to a level of 11,000 x 10^9 barrels. This upper limit to the potential future availability of oil emerges from totalling contemporary Western world estimates for conventional oil (6000 x 10^9 barrels) and for unconventional oil (5000 x 10^9 barrels). It is thus less than what must currently be viewed as the low probability for an even larger resource base as indicated, for example, in Soviet studies of the question. This elimination of the highest estimates is paralleled at the low end of the range where the few estimates[2] of total resources of less than 2000 x 10^9 barrels are also discounted.

There is, however, one other reason for not simulating a future of oil based on an estimate of resource availability higher than 11,000 x 10^9 barrels. This is related to the fact, as will be demonstrated in Chapter 5, that with a higher resource base there is no constraint on the future development of oil supply within the study period of 100 years, up to the year 2080, except in the cases of very high, and hence very unlikely, rates of

1. See A.Volkenborn (Maraven, S.A. of Venezuela), 'Venezuela's Heavy Oil Development Prospects and Plans', UNITAR Conference on *Long Term Energy Resources*, Montreal, November-December 1979 for a discussion of this process in respect of oil from the Orinoco oil belt.
2. See the delphi-type survey in the 1978 World Energy Conference, *op.cit.*, p.24.

increase in demand/or with severe constraints on the maximum annual rate at which the available oil can be proven. Larger figures of potential resources are thus just not worth modelling.

C. The Annual Rate of Additions to Proven Reserves

The second major component in the basic complexity surrounding the future of oil is that of the annual rate of additions to reserves. This, too, is a component in any consideration of the future of oil about which the post-1950 history of the industry has much to tell us. The nature of the successive processes of oil exploration, reservoir appraisal and development of field production, mean that the size of any discovery remains uncertain for a lengthy period after it has been located. Sometimes the uncertainty lasts as long as 20 years or more. The initial discovery well reveals only very limited information about the field and the declaration of reserves thus has to be small.[1] If, for one reason or another, that field is not appraised and/or developed immediately, then its proven reserves remain declared at a minimum initial level. With the appraisal of a field (by additional wells designed to try to define the amount of oil in place and the percentage oil in place which is likely to be recoverable) followed by its development (for production), the reserves of the field automatically increase[2] and so provide the basis for the process of reserves appreciation. This is the process in which the continued development of a field over time (that is, the 'depletion' of the reserves of a field as a consequence of production) serves to increase the size of the reserves to levels higher than those on which the initial development decision was taken. Paradoxically, therefore, the act of producing a field usually increases rather than decreases the size of the remaining reserves in the field. This sometimes applies for only a few years but quite frequently the phenomenon persists for many years during the production phase unless, in the meantime (as often happens) the rate of production from the field is increased

1. Indeed, in some countries/provinces the reserves' declaration has, by legislation, to be limited to the quantity of oil which can be produced by the discovery well. In all fields, other than the very smallest, this implies a declaration of reserves which is well below the real size of the field.
2. Except, of course, in the very limited number of cases when the initial expectations about the discovery from the exploration well – and these expectations underlie the appraisal and development decisions – turn out to be incorrect, either in terms of the quantity of oil in the reservoir or in terms of its producibility.

beyond the initially specified level in recognition of the greater opportunities offered by the larger-than-originally-expected reserves of the field. This phenomenon occurs without any concurrent changes in the economic environment – in terms, that is, of changed production costs and/or oil price levels. As improved production technology or a higher degree of recovery of the oil in place lead to lower unit production costs, more oil can be produced economically from a field. A similar result will emerge from an increase in the oil price as this makes it worthwhile to put more investment into a field's production system thereby enhancing the level of recovery. Changes in the economic environment will, moreover, enable a reappraisal to be made of an earlier 'no-go' decision in respect of a discovered oilfield. If development now becomes economic, as a result of higher prices then the earlier limited declared reserves (if any) will automatically be enhanced.[1]

Such economic aspects of reserves appreciation were, of course, of limited importance in the 20-year period of declining real oil prices, from 1950 to 1970, as shown in Figure 1.2. At that time the possibilities for appreciation were restricted to those arising from cost reductions through technological improvements. In today's conditions of rising real oil prices, however, the reappraisal of earlier sub-economic finds must clearly become much more attractive. Given time, one may now anticipate the appreciation of the world's proven oil reserves based on the economic, as well as on the technical component in the process.

The development and application of new technology contributes powerfully to the appreciation of reserves through the upward modification of the percentage of the oil in place which can be produced. An excellent example of this can be seen in the case of the Schoonebeek field in the Netherlands. This field was discovered in 1937 and was then, in the light of existing technology of heavy oil production, expected to produce no more than 5 to 15 per cent of the 2 x 10^9 barrels of oil estimated to be in place. Over 20 years later, however, in the early 1960s an experimental steam-drive system for enhanced recovery was installed on a part of the field. Ten years later its effect was carefully evaluated. The recovery factor had been

1. These, and other, issues relating to the economics of different levels of oil production from a field were discussed at length in P.R.Odell and K.E.Rosing, *Optimal Development of the North Sea Oilfields*, Kogan Page, London, 1976.

increased to 50 per cent or higher and, following this favourable result, the steam-drive technology is now being extended to the entire field. This minimum 3·3 to the maximum 10-fold appreciation of the field's reserves now makes Schoonebeek a giant oilfield instead of a relatively minor one.[1]

The ability to apply such technology to increase the recovery of oil from known fields depends in the first instance on economic considerations (basically, the relationship between the increased expenditure necessary and the higher revenues from rising oil prices). It also depends on the availability of the necessary technical and managerial expertise whereby the successful installation of such schemes is made possible. This latter factor will make the process a slow one, given the concentration of most of the world's oil producing capacity in countries where such expertise is in short supply and from which the international oil companies –with their wealth of expertise – have recently been excluded through nationalization measures. Nevertheless, the developments may eventually be widely applied and ultimately will have a quite formidable impact on the overall world-wide oil reserves' situation.

Conventionally, estimates of the ultimate world oil resource base are closely related to the historical record of the industry in recovering only a relatively small part of the total oil in place. This currently averages about 30 per cent. Each one per cent increase in the recovery rate enhances the world's reserves by the equivalent of well over one year's use at present rates of consumption.[2] Thus, an average increase world-wide to 40 per cent recovery would give another 10 years supply of oil – even with an annual rate of increase in use as high as seven per cent. Even better recovery rates can be expected from fields yet to be discovered (because they can be developed using the latest and most effective primary, secondary and tertiary recovery systems) so that the minimum 1000×10^9 barrels of oil from new fields included in the estimates of world oil resources, based on an assumption of the historic rate of recovery, could well be

1. See K.E.Rosing, 'Heavy Oil in Western Europe: a Survey' in: *The Future of Heavy Crude and Tarsands*, (R.F.Meyer, Ed.), Pergamon Press, New York, 1980 for a more complete discussion.
2. On the assumption that the figure of 1050×10^9 barrels of used plus presently proven oil reserves emerge from an expected average recovery of about 30 per cent of the oil in place. A one per cent improvement in recovery rates would give an additional 35×10^9 barrels of oil. World oil use in 1979 was about 23.6×10^9 barrels.

increased by between 60 and 100 per cent (assuming future recovery rates of 48 to 60 per cent instead of the historic 30 per cent).

The appreciation of oil reserves in the future thus seems likely to be high – based on the impact of a combination of economic and technological factors. Even historically, when some of these factors did not apply, appreciation has been significant in increasing reserves above the levels which were thought to exist. The impact of the appreciation of reserves on the evolution of the size of the world's proven reserves from 1940 to 1970 is shown in Figure 1.3. The solid line in this diagram shows British Petroleum's 1978 estimate of remaining, proven, non-communist world reserves for each year between 1940 and 1970[1] with the reserves back-dated to the year on which each field was discovered. The lower, dashed line shows the size of the proven reserves as they were declared to be at the end of each year. The difference between the values for each year is a measure of the degree of reserves' appreciation over the period up to 1978 and shows just how formidable this process has been in terms of increasing the historically recorded reserves' figures. For example, the less than 100 x 10^9 barrels of declared reserves in 1950 have increased to over 350 x 10^9 barrels and the 300 x 10^9 barrels declared to exist in 1960 are now known to have been under-stated by over 65 per cent.

Appreciation has been an equally formidable influence in enabling the reserves/production ratio for each year to be changed retrospectively. This is shown in Table 1.2 in which the R/P ratios for the period 1950-70, as calculated by BP in 1978[2], are compared with the ratios which were declared contemporaneously and which formed the data set on the basis of which the future of the oil industry was viewed from year to year. The third column in Table 1.2 then converts each contemporary nominal R/P ratio into an effective ratio by taking into account the average per annum rate of increase in oil use over the

1. There is, of course, absolutely no point in showing the situation post-1970 – over the period of the last 10 years, that is. This is because fields discovered since then are still either undeveloped or in the very early stages of production (as, for example, with the earliest North Sea fields) so that no realistic picture can yet have emerged of the size of the field. Even fields discovered between 1960 and 1970 and since developed are still likely to be appreciating to some degree while many non-economic finds made in that period will now be worth further appraisal for possible development. The future minimum likely appreciation of the reserves discovered between 1960 and 1970 is shown in Figure 1.3 by the topmost (dotted) line on the diagram.
2. See British Petroleum's publication, *Oil Crisis . . . Again?*, London, 1979.

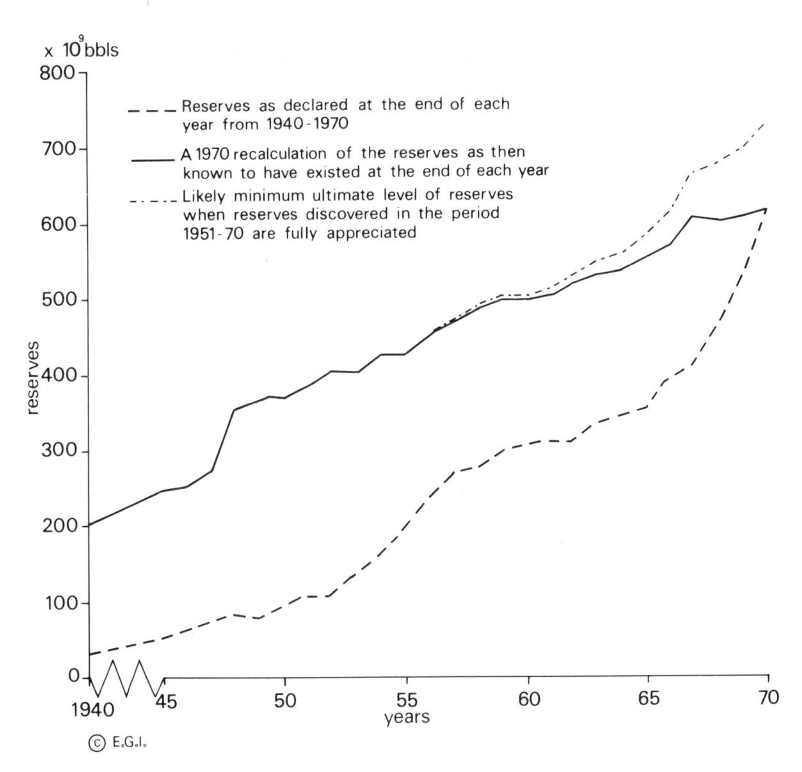

Figure 1.3 *Contemporary and Retrospective Estimates of Proven Oil Reserves*

previous 10 years. By comparing this historic set of effective R/P ratios with the present reserves' position it is clear that the future of oil, as reflected in the known proven world reserves situation, has seldom been as good as it is now. The present nominal R/P ratio stands at a level of about 27 years – somewhat lower than that of the nominal ratios as presented in the 1960s though better than in any year between 1940 and 1953. However, expectations now are, at best, for a slowly rising increase in oil use. This means that the effective R/P ratio is not very much smaller than the nominal one. Thus the present effective ratio of 25.8 exceeds the effective ratio of known reserves to future demand when compared with all other years except five (1957-61) in the last 30 years' history of the industry.

The reality of the oil reserves position at any specific moment is thus more complex than a simplistic presentation – such as

Table 1.2 *World proven oil reserves/production ratios, 1950-70 and 1979*

Year	1978 Calculation with appreciated Reserves (years)	Contemporary Calculation from Reserves as declared (years)	Effective Ratio allowing for increasing oil use	
			Effective Ratio (years)	Rate of Increase in oil use*
1950	95	24.7	19.2	5.41%
1951	91	23.8	17.6	6.50%
1952	90	25.9	18.9	6.74%
1953	84	27.8	20.3	6.76%
1954	84	31.3	22.7	6.97%
1955	75	33.5	23.4	7.77%
1956	75	37.6	25.6	8.28%
1957	73	40.4	27.7	8.13%
1958	75	41.6	30.2	6.95%
1959	71	41.0	28.7	7.71%
1960	66†	38.9	27.6	7.38%
1961	64†	38.0	28.1	6.55%
1962	63†	35.3	25.5	7.08%
1963	58†	34.7	25.0	7.08%
1964	55†	33.0	23.3	7.51%
1965	53†	32.0	23.2	6.94%
1966	51†	32.0	23.3	6.88%
1967	53†	31.7	23.0	6.98%
1968	46†	30.6	20.7	8.37%
1969	46†	34.0	23.8	9.75%
1970	45†	32.4	22.4	7.96%
1979	will not be known until 1999	27.4	25.8	1.32%**

Notes:

* Average annual rate of increase in oil use over previous 10 years shown in last column.

† These R/P ratios are likely to be under-stated as only a conservative view has been taken of the post-1970 appreciation of these reserves.

** Average annual rate of increase in oil use 1973 to 1978.

that of BP[1] – would appear to indicate. It is, however, very important to understand this complexity since it contains an element which is crucial for an effective interpretation of the future of oil. This is the matter of the annual addition to reserves required to maintain a lowest acceptable reserves to production ratio.

It is generally accepted in the oil industry that the effective forward planning of the industry's infrastructure necessitates proven reserves which are equal to the total amount of oil

1. *ibid.*

expected to be used in the following 10 years. If this were to be defined through the nominal R/P ratio then the required 10 year R/P ratio would be a function of the expected rate of increase in the use of oil. This, of course, rises increasingly steeply with higher rates of rising demand.[1] For the average rate of increase in oil use between 1950 and 1970 of about 7½ per cent per annum the required R/P ratio had to be almost 15 years. Reference back to Figure 1.3 shows that this was achieved throughout the 20-year period during which there was such a rapid rate of increase in the use of oil. Since 1973, however, the rate of increase in the demand for oil has fallen dramatically – to a rate which, in the six years since then, has only just exceeded 1.3 per cent. At this lower rate of increasing use 10 years' availability of oil is achieved through an R/P ratio of 10·6. This is a requirement which is significantly lower than the current unadjusted reserves to production ratio of over 27 years.[2]

The immediate future of oil thus clearly does not depend upon the continuing discovery of enough oil each year to match the current world total annual use of some $23{\cdot}6 \times 10^9$ barrels. Indeed, with a continuing growth rate in oil use of 1·32 per cent per annum (the 1973-8 average annual growth rate) it would be 1995 before the adjusted R/P ratio was low enough to require new oil discoveries – even assuming, completely unrealistically, that there was no appreciation whatsoever of currently known reserves. With a rate of appreciation no higher than that in the past, it would be 2002 before any new oil discoveries had to be made.

This favourable situation over oil reserves relative to demand, gives the oil industry a considerable breathing space in which to organize itself for the large-scale exploration and discovery of oil in new geographical locations (notably in the Third World and offshore) and/or in deeper formations, and/or from new habitats such as the tar sands and the oil shales. This is an important element in terms of the world outlook for oil, given the transition period through which the oil industry of the

1. The data and the derivation of R/P ratios are discussed in Chapter 2. See pp. 53-54.
2. This may be defined as 'unadjusted', given the certainty of appreciation of the reserves discovered to date – following the arguments set out above on pp. 36-37. Even if the historic trend of the process of appreciation is no more than repeated this makes the current unadjusted ratio more than 40 years. However, as also argued above (p. 34), changed economic circumstances seem likely to make appreciation even more important in the future than it was in the period of decreasing real oil prices between 1950 and 1970.

Western world is going in respect of ownership and organization. As a consequence of disturbing factors in this transition the Western world's industry as a whole is currently in a much less strong position to find new reserves, than it was during the period between 1950 and 1974. At that time the large international oil companies dominated the exploration and development efforts outside North America and were able to put their considerable resources of know-how and capital accumulation to work, without having to worry too much about international and national political considerations. The relative decline in the importance of these historically significant institutions in the search for, and the development of, oil resources, as well as the lack of ability and/or motivation on the part of many of the newly formed oil entities to continue with the oil finding and developing processes,[1] seems likely to have been the main factor which has restricted growth in non-communist oil reserves in the last few years. This is, almost certainly, a temporary phenomenon as most of the member countries of OPEC will eventually want to search for and to prove new reserves of oil.[2] Moreover, many previously oil-poor countries will ultimately be successful in their post-1973 oil crisis exploration efforts. Nevertheless, it seems appropriate to assume that, for the next decade at least, additional new oil plus appreciation of old oil will not, on average, be sufficient to replace fully the amount of oil which is used year by year. There will thus be a consequential continuing modest annual decline in the unadjusted R/P ratio.[3]

Indeed, in the immediate future the annual additional amounts of oil which have to be discovered or which must become available from appreciation need to be no more than modest. This is because all, or most of, the oil required to meet the now slowly increasing demand could come from the reserves that have already been discovered and appraised. From oil, that

1. This arises because they are mainly the state oil companies of the member countries of OPEC. Given the fact that most of these countries have rich existing known reserves, the state oil companies have little incentive to prove additional new oilfields or even to reappraise the volume of oil in oilfields which are already known.
2. Some OPEC countries have already taken the necessary action to achieve new discoveries and to reappraise old ones. These include Venezuela, Ecuador, Algeria, Nigeria and Indonesia where such developments are important to the countries' medium- to longer-term economic interest.
3. This is the way in which we have simulated the future of oil for the (variable) period post-1979 in which the nominal R/P ratio remains above the lowest level which is required to meet the next 10 years' oil consumption. See below, Chapters 2 and 5.

is, that is already 'on the shelf' as far as the industry is concerned. If one assumes, first, that the R/P ratio is steadily and continuously run down to the minimum necessary level and, second, that the use of oil continues to grow, then increasing quantities of oil will eventually have to be proved each year. It will be at this point in the continuing development of the industry that its ability to achieve a high enough annual discovery rate will become a critical variable in determining whether or not oil retains an expanding future.

There has recently been a great deal of speculation as to what the maximum annual finding rate for new oil might be. There have been suggestions that this rate may already have been achieved in the 1950s and the 1960s when the discovery of a series of super-giant fields[1] produced 'adjusted' additions to reserves of more than 100 x 10^9 barrels in a single year. These adjusted figures, however, involve the retroactive allocation of reserves to a particular year in the light of subsequent knowledge about the discoveries – and particularly the super-giant ones.[2] They were not the declared reserves for the fields concerned as recorded at the time from the data available. In the years between 1950 and 1979 such annual declarations of new and appreciated reserves ranged from a low of 13 x 10^9 barrels to a high of 74 x 10^9 barrels and there was a considerable degree of variation from year to year as a result of the specific incidence of particularly large discoveries and/or a company's or government's timing in declaring them. In order to 'spread' the impact of these random events a five-year running mean of annual discovery rates has been calculated and is shown in Figure 1.4. This shows an upward trend in the rate of additions to reserves in the period 1960-70 with an eventual highest figure of 55 x 10^9 barrels. More recently the figure has fallen back to around 25 x 10^9 barrels, but as explained above (see p. 40), this appears to be due to the changes in the ownership and organizational structure of the industry since 1973, together with the fact that the severe retrenchment in the rate of increase in the demand for oil has undermined the earlier confidence of the industry in what it had hitherto come to accept as a process of inevitable expansion. Thus large annual additions to reserves

1. Internationally, super-giant fields are defined as those having at least 5 x 10^9 barrels of recoverable oil. In North America the same adjective is used to describe fields with more than 2 x 10^9 barrels of recoverable oil.
2. This is, for example, the approach used by British Petroleum for its publication, *Oil Crisis . . . Again?*, op.cit., p.4.

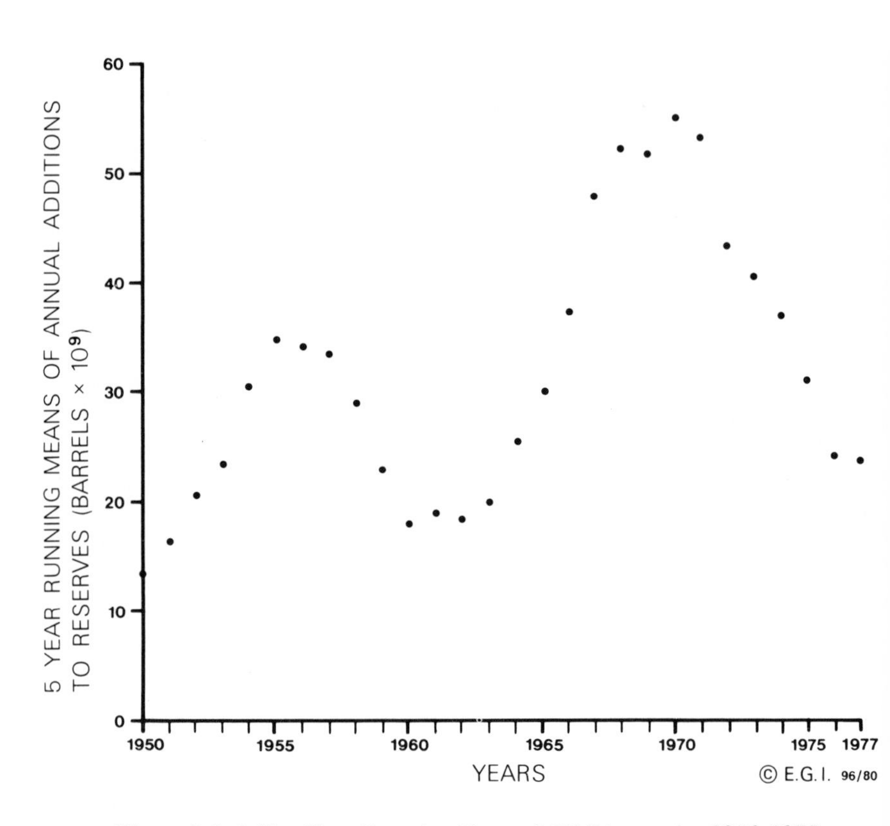

Figure 1.4 *A Five-Year Running Mean of Oil Discoveries 1950-1977*
(Based on contemporary declarations of proved reserves, i.e. with no allowance made for the later appreciation of reserves.)

have become less important.

It is difficult, therefore, to judge what the maximum annual rate of additions to reserves will eventually turn out to be. Given the vast areas with petroliferous potential which have so far remained untouched or nearly untouched by exploration,[1] and given the now increasing motivation on the part of most countries to stimulate the process (as a result of the high price of imported oil), there seems to be no reason why higher future annual additions to reserves cannot be achieved. It is also possible that the importance of the issue will, in any case, soon be severely diminished in view of the likelihood that reserves of non-conventional oil from tar sands and oil shales can be incorporated in the resource base. These non-conventional oil resources will be known to exist in vast quantities as a result of the direct geological interpretation of the data which becomes available. The process of gradually 'proving' the reserves with exploration, appraisal and development wells – as in the case of conventional oil the reserves of which can be proved only by the drill – is not required.[2] This latter possibility suggests that the future of oil can be modelled without a maximum annual finding rate constraint. More pessimistically, annual additions-to-reserves constraints, related to what the industry has already achieved and in the light of the expected size of the resource base, must then be introduced in order to give an alternative view of the future of oil. This must be done in case non-conventional oils only become available on a large scale at a much later period in the development of the industry.

It is apparent that the maximum annual rate of additions to reserves must bear some relationship to the size of the resource base. If more than half the world's ultimate reserves of oil has already been found (as would be the case given the discovery to date of a total of some 1100 x 10^9 barrels and if the resource base is only 2000 x 10^9 barrels), then higher annual finding rates

1. These characteristics of the world-wide occurrence of and search for petroleum resources are illustrated in Figures 1.5 and 1.6. See also P.R.Odell, *World Energy, Needs and Resources*, Institute of Bankers, London, 1979, for a more detailed discussion of these considerations.
2. Oil reserves in such non-conventional forms have much more in common in respect of their reserves' status with coal than with conventional oil. The reserves are so large relative to production that the concept of an R/P ratio becomes ludicrous. Reserves' development is now essentially a function of economic conditions and, of course, of the availability of the technology; first, generally in terms of its evolution and second, in terms of its specific application to a particular country or region.

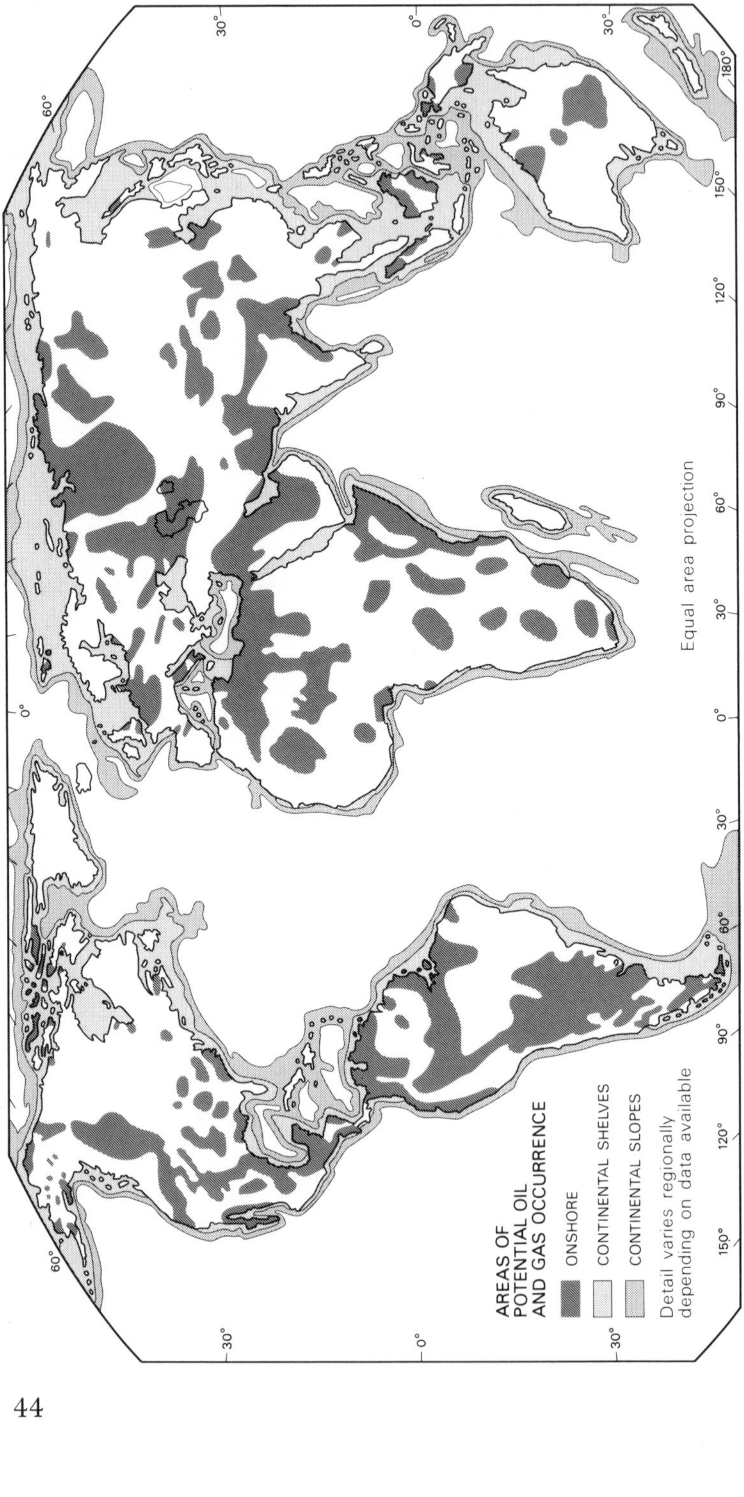

AREAS OF POTENTIAL OIL AND GAS OCCURRENCE

ONSHORE

CONTINENTAL SHELVES

CONTINENTAL SLOPES

Detail varies regionally depending on data available

Equal area projection

© E.G.I. 114/80

Figure 1.5 *The World's Potentially Petroliferous Regions*

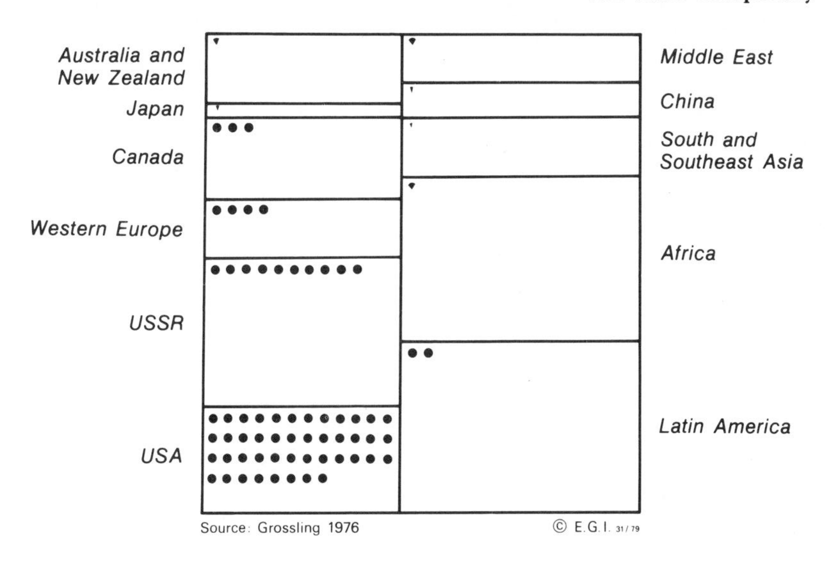

Figure 1.6 *The Regional Distribution of the World's Potentially Petroliferous Areas, Shown in Proportion to the World Total*

Within each region the number of exploration and development wells which have been drilled is shown — each full circle represents 50,000 wells. Relative to the US, all other parts of the world — but especially the regions of the Third World — are little drilled for their oil.

than those already achieved are most unlikely. This is because more and more effort will have to go into locating oil in increasingly difficult locations and habitats so that the productivity of exploration investment is bound to fall significantly and more or less continuously. On the other hand, if only just over 20 per cent of the world's ultimate conventional oil has been found to date (as in the case of a resource base of 5000 x 10^9 barrels) then, unless finding rates increase markedly above those achieved to date, the oil in the ground is going to take well over another 100 years to locate and to prove. This is clearly very unlikely unless one assumes that technology will not develop, or that it will actually deteriorate from the levels which have already been achieved. Alternatively, one would have to make the assumption that the rate of growth in the use of oil will be so slow that the motivation to find more oil more quickly is never realized. In other words the maximum annual additions to reserves variable is also one to which it is necessary to attach a wide range of values in any comprehensive look at the future of oil. This has been done in this study: the range used and the method of incorporating the variable into the model are specified in Chapter 2.

D. The Use of Oil

The third main component in the future of oil is the evolution of the use of the commodity. In contrast with the other two main variables, however, this aspect of the outlook for oil cannot easily be related to the industry's experience in the period from the early 1950s to 1973. During that period the use of oil moved ahead at an exponential growth rate of over 7 per cent per annum under the combined stimulus of rapid economic expansion in most parts of the world, a declining real price for oil (see Figure 1.2) which encouraged increasing energy-intensive ways of doing things and of going places, and the substitution of coal and other sources of energy by oil in many parts of the world. Table 1.3 indicates the results of these factors in increasing the use of oil in a selection of industrialized countries[1] whilst in Table 1.4 the 1950-75 changes in energy/oil use in Latin America (as an example of changes in the Third World) are set out.[2]

Between 1974 and 1979 both the general economic and the more specific energy situations have changed quite dramatically so that the evolution in the use of oil, over this recent period, bears no relationship to the earlier experience of the industry. World-wide, between 1973 and 1978, the annual rate of growth in the use of oil was only 1·32 per cent and if the communist countries are excluded (their economies have been less seriously affected by the traumas of the 1970s) then the rate of increase in oil use over the period falls to a figure as low as 0·65 per cent per annum.

It is highly inappropriate to think in terms of the evolution of the future use of oil as being anything like the evolution of the demand curve in the period up to 1973. The reasons are first, the continuation of oil price increases (the level of real oil

1. This Table is based on one from P.R.Odell's *The West European Economy*, Stenfert Kroese, Leiden, 1976, in which the factors affecting developments in oil use in Western Europe are discussed at length.
2. This is from P.R.Odell, *Latin America's Energy Prospects*, EGI Working Papers Series A No.79-16, Rotterdam, 1979, in which the evolution of the energy/oil market in Latin America is also discussed. A shortened version of this paper (without the Tables and other illustrations) has been published in the *Bank of London and South America Review*, May 1980.

OFFSHORE PETROLEUM ENGINEERING

A Bibliographic Guide to Publications & Information Sources

Marjorie Chryssostomidis, Ocean Engineering Librarian, James Madison Barker Engineering Library, Massachusetts Institute of Technology

This guide consists primarily of a bibliography citing some 2,600 books, conferences and reports on offshore petroleum and related topics. The citations are arranged by subject area with the necessary bibliographic information for ordering each publication and a descriptive abstract for most of the books and conferences. Author and title indexes are included as well as a directory providing addresses of publishers and other organizations.

Contents

A NEW KOGAN PAGE ENERGY PUBLICATION

BRITISH OIL POLICY: A RADICAL ALTERNATIVE

Peter R Odell

This study takes a critical view of the legal, financial and regulatory framework which has evolved since 1964 for the exploitation of Britain's offshore oil resources. It also seeks to examine whether the decisions which are being taken by the government are based on a true picture of the future world energy situation. Peter Odell presents an alternative view of the future of oil and suggests a different basis for the exploitation of resources. The study also demonstrates why it is necessary to integrate the production of oil into the country's pre-existing oil refining and distribution activities in order to ensure that the benefits from the exploitation of Britain's resources are maximised.

£12.00, Paperback ISBN 0 85038 316 1
A4 (297x210mm) 100 pages approx

UNITED KINGDOM OFFSHORE LEGISLATION

A NEW KOGAN PAGE ENERGY PUBLICATION

Table 1.3 *Energy use in selected industrialized countries, 1952-72*

Year	Energy used	Country					
		US	*United Kingdom*	*West Germany*	*France*	*Italy*	*The Netherlands*
1952	Total mtce*	1271	232	145	89	25	22
	Tons/capita	8.2	4.6	2.9	2.1	0.6	2.1
	% Coal	34	90	95	79	43	78
	% Oil	43	10	4	18	35	12
	% Gas	22	–	–	–	8	–
	% Other†	1	–	1	3	15	–
1957	Total mtce*	1460	247	186	111	44	28
	Tons/capita	8.8	4.8	3.5	2.5	0.9	2.5
	% Coal	27	85	88	72	28	63
	% Oil	45	15	11	25	48	36
	% Gas	26	–	–	1	15	1
	% Other†	2	–	1	3	9	–
1962	Total mtce*	1646	265	221	122	71	35
	Tons/capita	8.9	4.9	3.9	2.6	1.4	2.6
	% Coal	23	72	71	58	18	47
	% Oil	41	28	27	33	62	51
	% Gas	34	–	1	5	13	2
	% Other†	2	–	1	4	8	–
1967	Total mtce*	1963	276	251	154	112	47
	Tons/capita	10.0	5.0	4.2	3.1	2.1	3.7
	% Coal	22	59	51	40	12	25
	% Oil	41	39	46	50	71	53
	% Gas	36	1	3	6	11	22
	% Other†	1	1	1	4	6	–
1972	Total mtce*	2428	302	333	215	152	76
	Tons/capita	11.6	5.4	5.4	4.2	2.8	5.7
	% Coal	20	40	35	21	7	6
	% Oil	44	46	52	66	75	36
	% Gas	34	12	11	10	14	59
	% Other†	2	2	2	4	4	–

Source: Based on UN Energy Statistics, Series J.
Notes:
* millions tons coal equivalent.
† mainly hydro-electricity converted to coal equivalent on heat value basis of the output. 1972 figures also include some nuclear power converted on the same basis.
Most important energy source in each country for each year is underlined.

prices doubled again from 1978-9 to cause a leap in the most recent part of the curve on Figure 1.2 which is even more dramatic than that of the 1973-4 period); second, continuing economic difficulties in the whole of the Western world; and third, the impact of energy conservation policies in general, and of oil substitution policies in particular, in all the non-communist

Table 1.4 *Energy use in Latin America, 1950 and 1975*

	1950		1975	
	mtoe*	%	mtoe*	%
Total energy use	*68.0*	*100*	*231.1*	*100*
of which				
Vegetable fuels	29.2	42.9	41.4	17.9
Oil	28.8	42.4	134.4	58.2
Natural gas	2.9	4.3	33.7	14.6
Coal	5.5	8.1	11.0	4.8
Hydro-electricity†	1.6	2.4	10.6	4.6

Notes:
* mtoe = million tons of oil equivalent.
† Calculated on basis of heat value of electricity produced 1kWh = 3412 btu = 860 kcals.

world's major energy consuming countries.

Gross uncertainties in the Western world's system – both economic and political – plus technological changes in energy using systems, together with societal reaction to much higher oil prices, are certain to make the future rate of increase in the use of oil much lower than in the period up to 1973. Instead, there will be, generally, a much lower rate of increase in the use of oil and within this general pattern of evolution of oil use there seems likely to be accentuated temporal and spatial variations in the growth rates. These variations will be due to contrasting economic fortunes in different countries and regions and to the unevenness with which energy/oil conservation policies are pursued and implemented by governments, either through the use of the pricing mechanism or by more directly interventionist and/or regulatory measures.[1] In these uncertain circumstances it is clearly necessary to use a wide range of oil use growth rates in a simulation study of the future of the industry. The fact that this has not been done in recent studies on the future of oil[2] reflects first, an apparent belief that a 10-year doubling rate in its use (as was achieved between 1950

1. See P.R.Odell, *The Western European Energy Economy*, and *The World's Energy Needs and Resources*, op.cit., for a more detailed discussion of these and related issues.
2. See, for example, OECD, *World Energy Outlook*, Paris, 1977; MIT Workshop on Alternative Energy Strategies, *Energy Global Prospect 1985-2000*, New York, 1978; British Petroleum, *Oil Crisis . . . Again?*, London, 1979; and Shell, *The Outlook for Oil 1980-2020*, London, 1979.

and 1960 and between 1960 and 1970 and as was expected to continue thereafter) is an inevitable 'fact' of modern economic life; and second, an unrealistic view of the ability of the Western economic system speedily to recover its economic equilibrium and its earlier propensity for strong and continuing growth. This, in part, reflects the difficulties of politicians and of policy makers in coping with the implications of an enforced end to the revolution of rising expectations on the part of the populations of the industrialized countries. Thus, their forecasts of economic growth have to be higher than those justified by the realities of the changed international economic system. And as policy makers still accept, either explicitly or implicitly, the idea of a close link between economic growth and energy use, their forecasts of the latter are necessarily also higher than the changing conditions justify.[1] The consequence of the failure to recognize the very much more modest outlook for the future use of oil compared with the world's experience in recent decades is, of course, a serious over-statement both of the rate of depletion of the oil resource base and of the speed with which new reserves have to be found and proven. Demand considerations thus form an essential and an uncertain element in the basic complexity of the future of oil. In Chapter 5 of this study an appropriately broad range of demand growth rates is incorporated into the simulation of the future of oil in order to ensure that the component is given due weight in the overall evaluation of the options open to society in respect of its energy needs.

1. Such views also ignore the increasingly strong evidence that further economic growth in industrialized economies does not necessitate much, or even any, increase in energy use for the next 20-40 years. This is because the conditions of the 1950s and the 1960s encouraged such a wasteful use of energy that the elimination of waste could, even in a growth economy, inhibit the need for additional energy use. See, for example, R.Stobaugh and D.Yergin, *Energy Future*, Random House, New York, 1979 and G.Leach et al, *Towards a Low Energy Strategy for the UK*, Scientific Publications, London, 1978 for analysis of this economy/energy relationship in the US and the UK respectively.

Chapter 2

The Structure of the Model

A. Introduction

In the previous chapter the basic concepts of the study were introduced. This chapter is intended to provide a verbal and mathematical description of the operation of these concepts which allows the calculation of quantitative estimates of the future of oil. Qualitatively the model is extremely simple: production goes up and then production comes down. The processes which account for this are first, the growth of the use of oil and second, the finite nature of oil resources.

With the increasing use of oil, production grows to meet demand until that time in the future when the volume of total ultimately recoverable reserves of oil can no longer support continued growth in production or, indeed, the level of production which has been achieved. At this point production begins to fall until, eventually, it declines to zero as all the world's oil resources are finally used up. There is, however, one other important constraint on this process. In order to maintain the rising annual level of production it is possible that more oil reserves would, at some time, have to be added to those reserves already known in a particular year or series of years. If this is the case then a more likely maximum level of additions to known reserves binds the model and effectively limits the peak production that can be achieved.

The relationship between the size of the resource base and the maximum possible additions to reserves in any one year has been discussed in Chapter 1. There (see pp. 43 to 46) the hypothesis that the larger the resource base the larger the maximum annual reserves additions is explained. In Figure 2.1 the curve defines our assumption of the relationship between the two variables and along the curve we have marked the

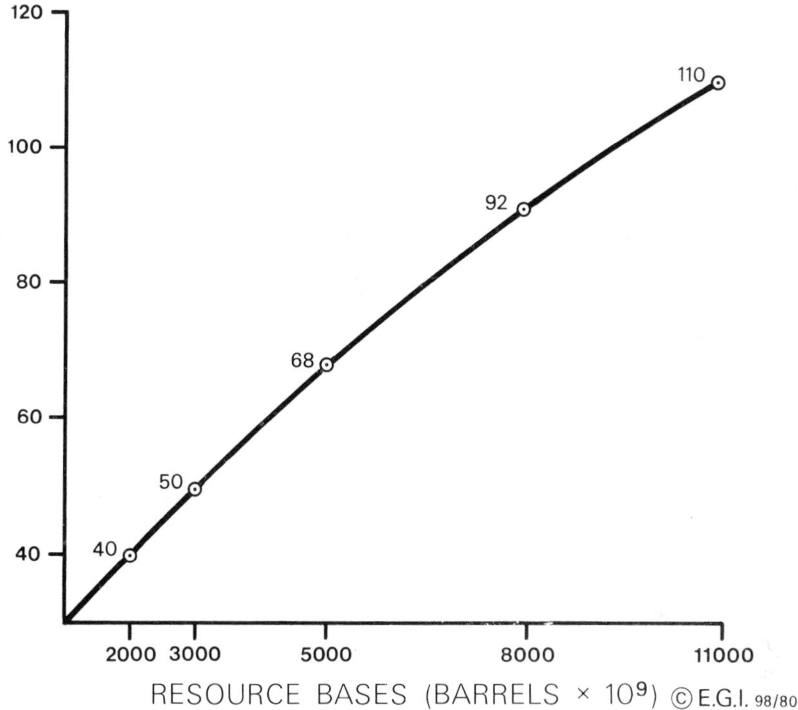

Figure 2.1 *Size of the Resource Base and Maximum Addition to Known Reserves Allowed in Limited Cases*

The numbers associated with the marked points on the curve are the limits used with the different sized resource bases.

specific assumptions made concerning the maximum possible annual additions to reserves for the five different sized resource bases with which we are concerned in this study.

B. The Model

Mathematically the construction of the model is complex. The three basic concepts of ultimate resources, annual additions to reserves and cumulative use are interrelated in that they affect or limit one another. The nature of these effects or limitations must be specified in the model as relational parameters, and this involves the introduction of a new series of assumptions. The structural elements of the model are specified in the following sections of this chapter (sections C-E) but it must be made clear that the model as programmed is even more complex than these specifications indicate. This additional complexity arises from the fact that all possible eventualities which could occur during over 100 different combinations of the parameters have had to be covered. In general, these complications occur because various checks, tests and calculations, included in the most relevant section of the model, also have to be incorporated in other sections of the model where they can, under certain circumstances, form alternative methods of calculation. Though these alternative methods of calculation rarely have to be used, as they apply only in cases of the most extreme combinations of parameters, when they are used they may cause entire sections of the model to be omitted. In order to cover these cases, calculations and tests had to be included in preceding or succeeding sections of the model so as to ensure the proper functioning of the program. These operationally necessary complications in the programming have been left out in the mathematical statement of the model in this chapter in order to avoid obscuring the important principles on which the model is based. Thus the description in this chapter is a sort of schematic presentation. Readers who wish to study every detail of the approach to the operation of the simulation of the future of oil are referred to the appendix to this chapter where the computer program version of the model is fully listed.

C. Definition of terms

Before proceeding to discuss the model we must define the terms which are used in the various mathematical expressions.

For convenience this is done in three sections; first, a description of the input parameters; second, a description of the four output vectors; and, lastly, the definiton of a series of time periods.

(i) *The Input Parameters*

The input parameters, taken together, define a particular case or run of the model. Two runs of the model in which any one (or more) of these parameters is (are) assigned a different value will have a different output and so be a different case. These parameters are, therefore, exogenous to the model and so define and control its functioning. They are, in general, constant in value throughout any one case.[1]

The first three parameters relate one by one to the three concepts discussed in Chapter 1. We define:

v = the total volume of ultimately recoverable reserves, that is, the resource base; (1)
f = the annual growth (or decline) rate of oil production/use; (2)
d = the maximum addition to the discovered reserves permitted in any one year. (3)

In order to calculate the required results from these three parameters certain other parameters which express elements of the relationships between them are also required. Define:

r = the reserves to production ratio; (4)
x = the proportion of v which can be found before the discovery of new reserves begins to constrain production; (5)
g = the rate of change of the discovery of new reserves once x is surpassed; (6)
m = a minimum level of production in cases with negative growth rates (decline rates). (7)

The first three parameters have already been discussed and explained in Chapter 1. The last four have not, however, been previously introduced and a brief discussion of each of them is required.

The reserves to production ratio (4) of any one year is given by dividing the total of all discovered and not yet produced reserves (= remaining proven reserves) by the year's production. It is a year-end calculation. In Chapter 1 this is referred to as the 'nominal' reserves to production (R/P) ratio. In order to have a

1. There is one important exception to this, in the case of the value of the growth in oil use parameter. In some iterations the defined value cannot be maintained throughout the growth period because of pressure on the resource base. In these cases the input value is substituted by a set of values generated by the model. See below pp. 71 and 73.

10 years' supply of oil available some number, larger or smaller than 10, must be used depending on whether production is increasing or decreasing and at what rate. The R/P ratio used in the model is, therefore, a function of f (2), the growth or decline rate of production. The nominal R/P ratio has to be adjusted to give an actual next 10 years' supply of oil given our assumption (explained in Chapter 1, see pp. 36 to 39) that this is a requirement of the industry for planning and infrastructure development purposes. With a positive growth rate in oil use, f (2), the nominal R/P ratio must be greater than 10 and with a negative growth (decline) rate, f (2), the R/P ratio will be less than 10. Figure 2.2 shows the relationship between f (2) and r (4) over the range of growth rates, f (2), from –1.0 per cent to +7.5 per cent per annum. The horizontal axis shows the various input values we use for f (2); *viz* 0.65, 1.32, 2.50, 3.30, 5.00, and 7.50 per cent. The vertical axis defines units of R/P ratios, r (4), and the values of r (4), which correspond to the input values of f (2), are marked along the curve.

The R/P ratio is a relational parameter operating between the annual level of production and the annual addition to the known reserves. As such, it is used to calculate the required annual addition to reserves when production is growing at the defined rate. It can, however, also be used to calculate the level of production if the addition to reserves, required in order to maintain the defined growth rate, cannot be maintained. In this latter case the R/P ratio parameter is used to calculate the production growth.

The parameter x (5) determines the point at which growth stops and decline begins. Once more than x (5) per cent of v (1) has been discovered, the method of calculation of the production curve undergoes a radical change. In the normal operation of the model as long as the cumulative discovered reserves are less than x (5) per cent of v (1), annual production grows by f (2) per cent each year. Conceptually, however, we would argue that once the cumulative discovered reserves are greater than x (5) per cent of v (1), then the rate at which new reserves can be added to already known reserves becomes constrained. There are two reasons. First, so much of the ulimate resource base is already discovered that the continuing exploration effort yields only a declining return (in terms of the volume of new reserves per unit of exploration effort), since more and more of the remaining undiscovered resources are in small and/or difficult to detect reservoirs. And second, such a high rate of recovery has

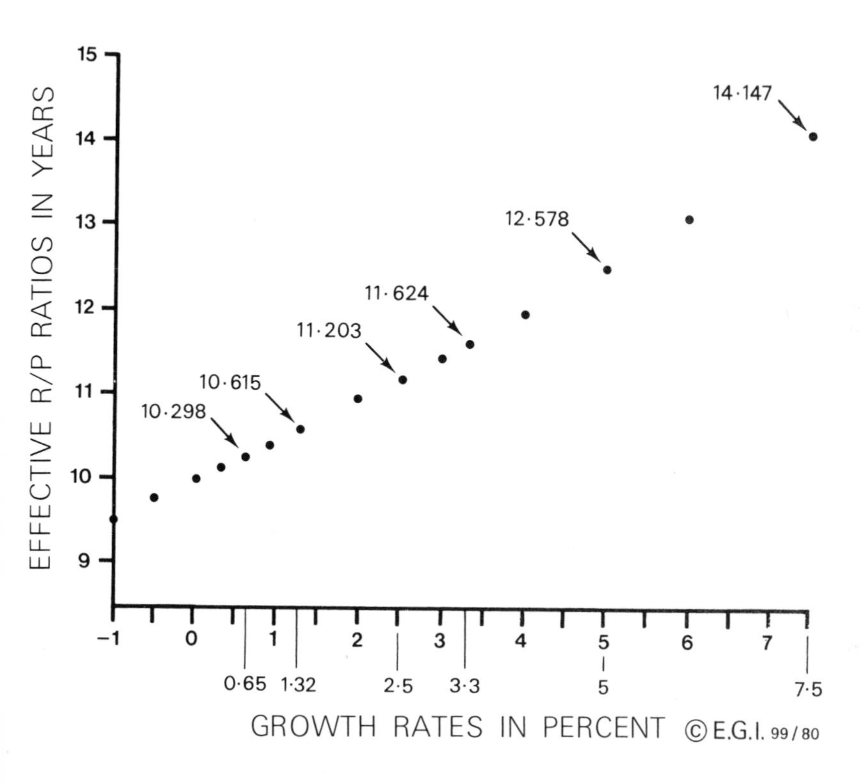

Figure 2.2 *Growth Rates and Effective R/P Ratios*

In order to maintain a proven supply of oil sufficient for the next 10 years the R/P ratio must be related to the growth rate in the use of oil. The six values identified on the curve are the R/P ratios associated with the values used for the growth rates in this study (related to a 10 year R/P ratio with a growth rate of zero).

already been achieved from known reservoirs that, adding additional reserves through the further improvement in the recovery factor, now requires very high and complicated technology. This implies considerable financial and time investments in research. Operationally, this constraint on reserves additions is modelled by reducing the annual addition to reserves by g (6) per cent per annum once x (5) per cent of v (1) has been discovered.

The effects of these two variables are clearly related and this means that a decision on an appropriate value for one of them will help to determine the appropriate value of the other. In order to decide appropriate values for the two variables a number of iterations of a particular case (*viz* 3000 x 10^9 barrels of ultimate resources and a growth rate in oil use of 1.32 per cent per annum) were repeatedly run using different values for x (5) and g (6). Figures 2.3 to 2.11 are included to demonstrate the effects of different values for these variables. For the sake of clarity we shall first consider the variables separately and then look at their combined influence.[1]

The variable x (5) is basically a timing variable as it separates the period of growth from the period of decline. At a particular moment after a year in which x (5) per cent of v (1) is reached, the production of oil must begin to decline. In Figure 2.3 a high value (75 per cent) of x (5) is tested. This value of x (5) is achieved in this iteration in the year 2023 and over the next 20 years the remaining 25 per cent of the world's oil resources is added to known reserves. In this instance the decline period is short and it starts from a level of production which has been allowed, by the high value of x (5), to a level that is too high by 2043 when all reserves are known. This necessitates a catastrophic fall in the rate of production as the limited remaining reserves in the post-peak production period are used up.

Figure 2.4 presents an iteration in which there is a less extreme value for x (5), *viz* 66.7 per cent. This proportion of the resource base is proven by 2016 but the remaining one-third of resources are added to known reserves over a period of no less than the next 42 years. This inhibits the rise of production to such a high level and allows a long enough period for the decline in production to occur at a reasonable rate.

1. Readers wishing to study these nine illustrations in detail at this stage will need first to turn to section F of this chapter in which there is an explanation of the manner of presentation.

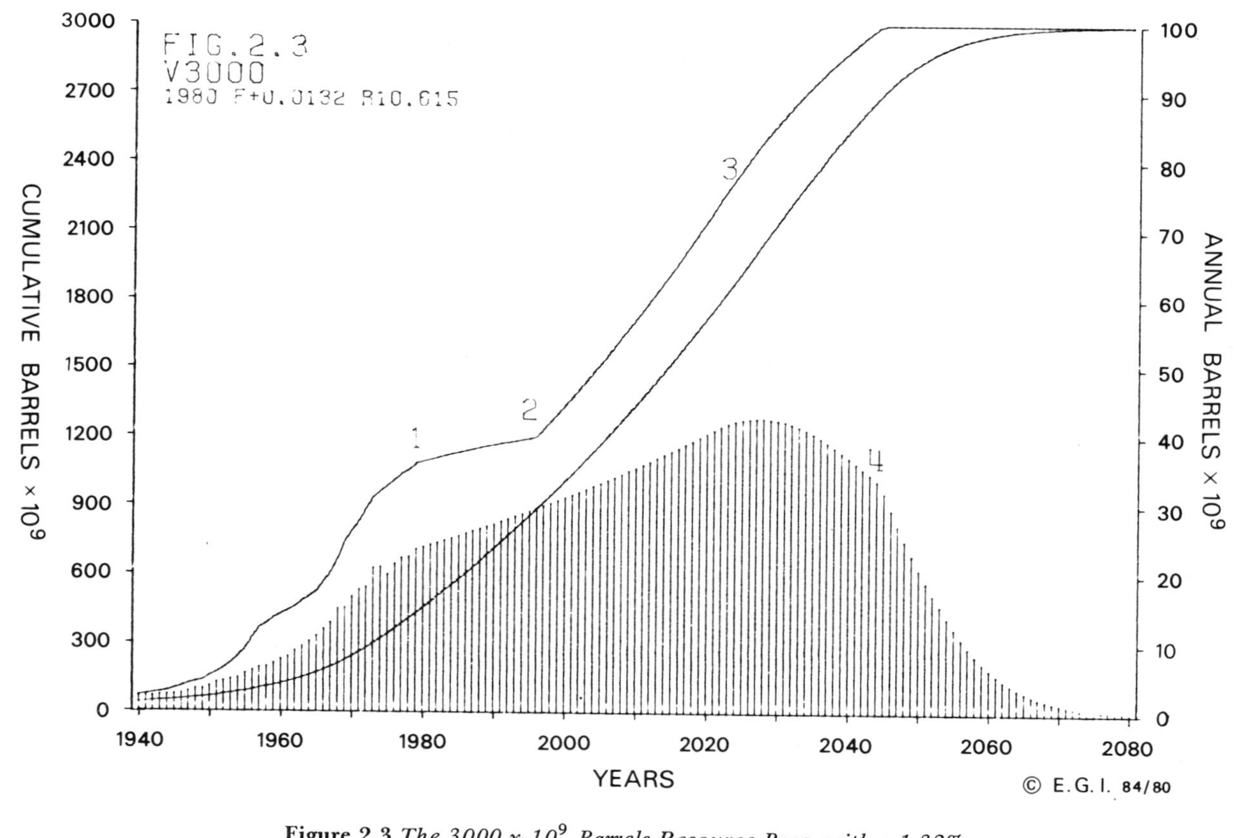

Figure 2.3 *The 3000 x 10^9 Barrels Resource Base, with a 1.32% Growth Rate Case, High x (75%), Moderate g (3%)*

For explanation of key on diagram see Figure 2.13 and accompanying text.

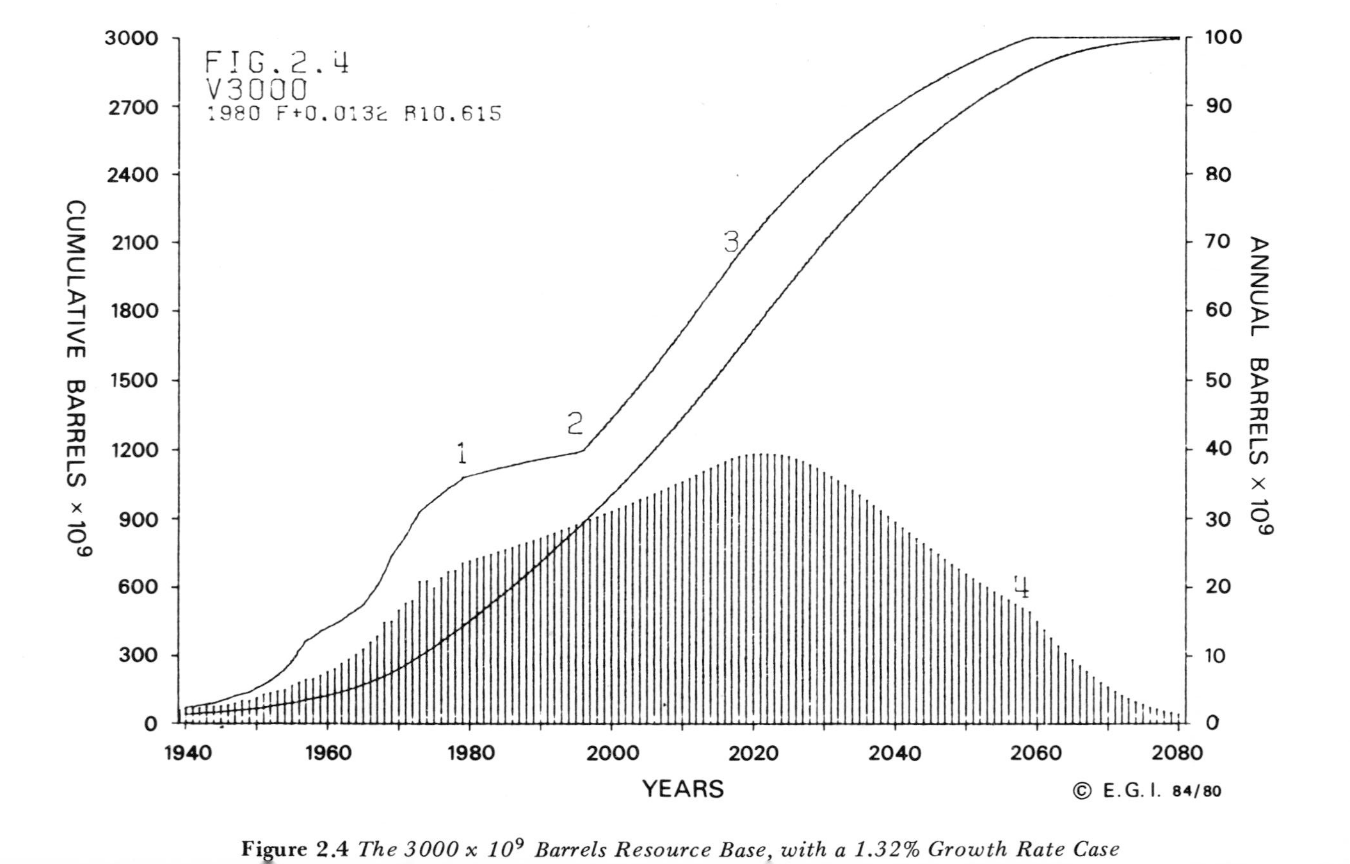

Figure 2.4 *The 3000 x 10^9 Barrels Resource Base, with a 1.32% Growth Rate Case*

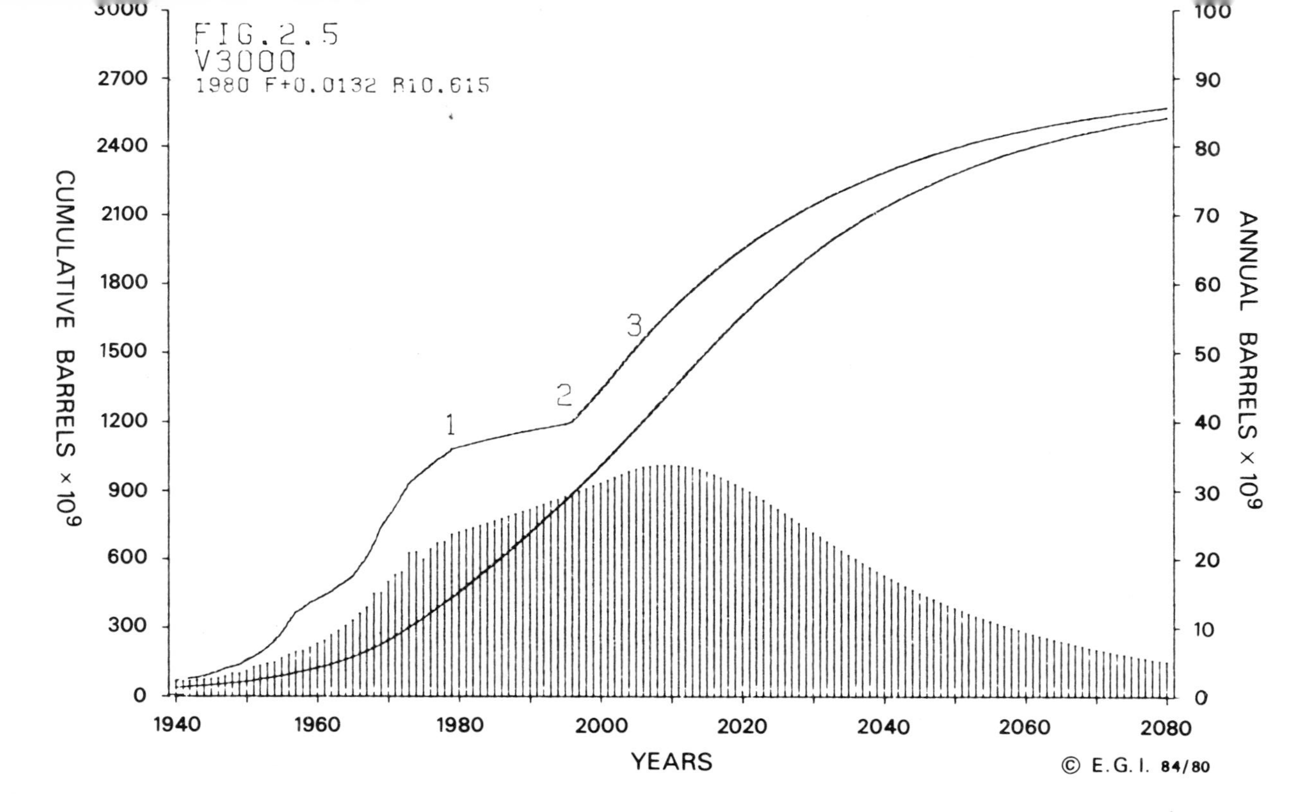

Figure 2.5 *The 3000 x 10^9 Barrels Resource Base, with a 1.32% Growth Rate Case Low x (50%), Moderate g (3%)*

An even lower value of x (5), *viz* 50 per cent, was also tested and is presented in Figure 2.5. The decline in production now has to begin so early that the annual volume of additions to reserves is asymptotically approaching zero by 2080 but, by then, with only 85 per cent of the total resources discovered. This is an unacceptable situation.

The variable g (6) controls the shape of the decline curve in the rate of annual additions to known reserves, once x (5) per cent of v (1) has been discovered. The rationale behind this is that at some point the increasingly small quantity of unproven reserves will become inherently more and more difficult to add to the known reserves. This is true both in respect of the difficulties in making new discoveries and the problems which grow over time in improving the recovery factor from known fields. The value of g (6) is thus important in defining the shape of the production decline curve.

Figure 2.6 shows a high value of g (6): a value of five per cent, that is, for the rate of decline in the annual volume of additions to known reserves. This, as the diagram shows, creates the same problem as that in the iteration presented in Figure 2.5, *viz* a near zero rate of growth in known reserves by the year 2080 even though about eight per cent of the ultimate resource base remains to be discovered.

A lower value (one per cent) for g (6) has been tested and is shown in Figure 2.7. In this case the catastrophic decline of the industry, once all reserves are known, is even more dramatic than that observed in respect of the case shown in Figure 2.3. The moderate case for g (6) is shown in Figure 2.4. Here the variable has a value of three per cent and a reasonably shaped depletion curve is produced.

In each of the cases presented in Figures 2.3 to 2.5, in which x (5) was varied, the value of g (6) was held constant at three per cent. In each of the cases where g (6) was varied (see Figures 2.4, 2.6 and 2.7), the value of x (5) was held constant at 66.7 per cent. The results of this series of cases clearly indicate that the combination of g (6) equal to three per cent and of x (5) equal to 66.7 per cent offers a preferable shape to the production curve. It avoids the drawbacks (such as too high a peak and too steep a decline) to which we have already alluded.

There are, however, still four other combinations of values of these two variables which have not yet been investigated. These are the cases in which x (5) and g (6) are varied together. They are illustrated in Figures 2.8 to 2.11.

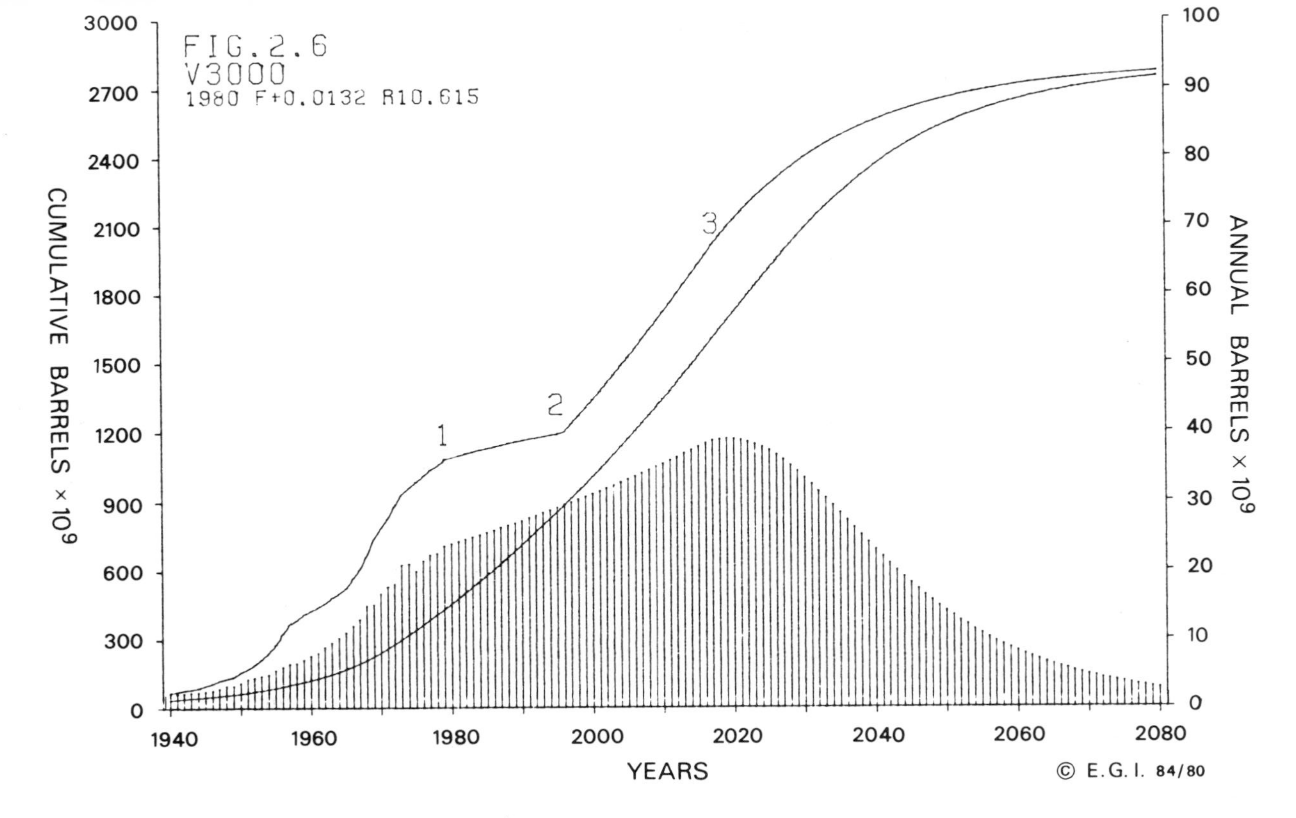

Figure 2.6 *The 3000 x 10^9 Barrels Resource Base, with a 1.32% Growth Rate Case Moderate x (66.7%), High g (5%)*

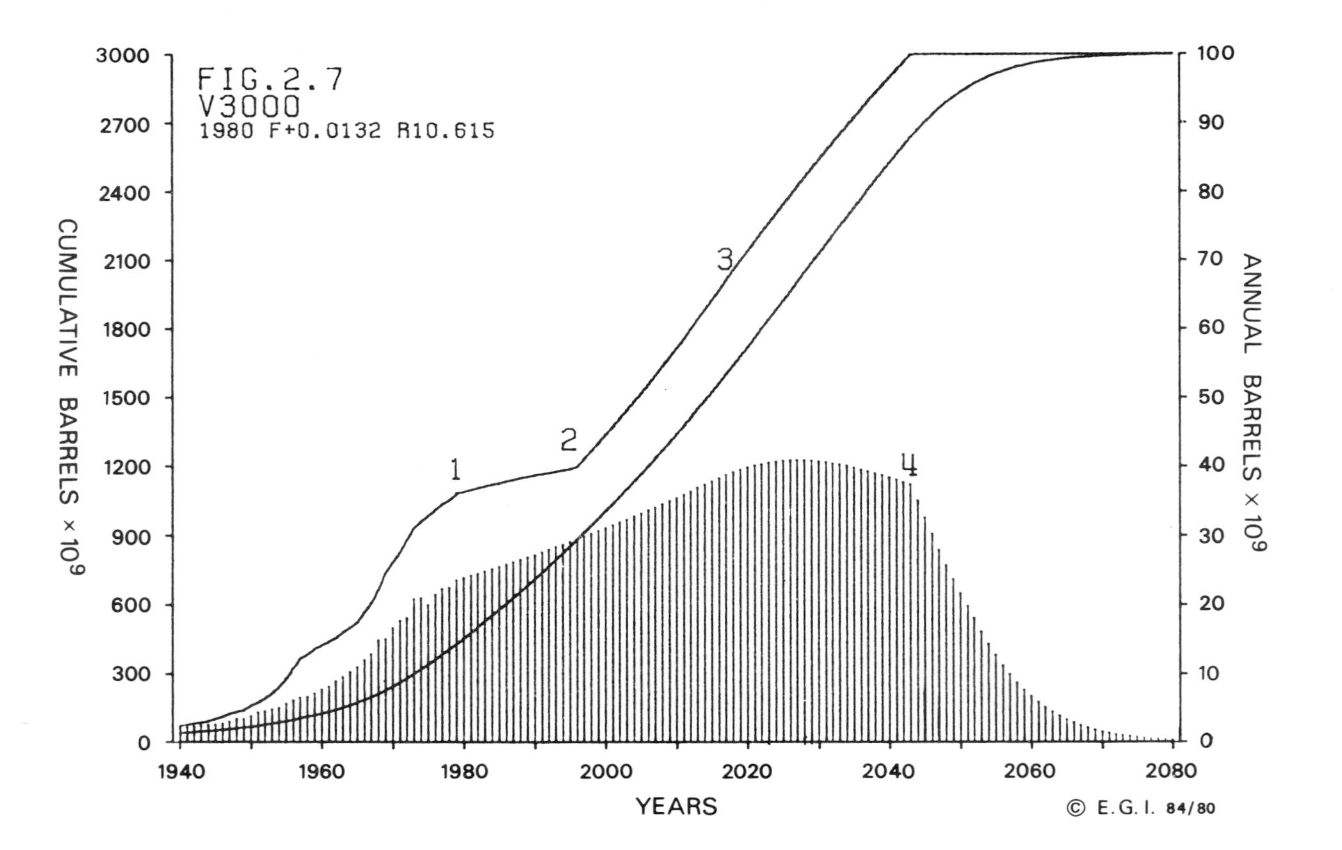
FIG.2.7
V3000
1980 F+0.0132 R10.615
CUMULATIVE BARRELS × 10^9
3000
2700
2400
2100
1800
1500
1200
900
600
300
0
ANNUAL BARRELS × 10^9
100
90
80
70
60
50
40
30
20
10
0
1940
1960
1980
2000
2020
2040
2060
2080
YEARS
1
2
3
4
© E.G.I. 84/80

In Figure 2.8, g (6) has a value of five per cent and x (5) has a value of 50 per cent. This is the low x (5)/high g (6) combination and it provides the worst example yet in terms of resources which remain undiscovered in 2080. At that time 27 per cent are still undiscovered although there is by that date a near zero rate of discoveries. Figure 2.9 still shows g (6) at five per cent but x (5) is now 75 per cent. This provides quite a reasonable result since the high value of g (6) compensates for the high value of x (5) giving a long, albeit relatively, steep decline of production but avoiding catastrophic collapse. Our preference for the combination values shown in Figure 2.4 is an indication of our judgement that the decline forced on the industry by resource based limitations will have to be a more gradual one than that of Figure 2.9.

In the remaining two cases (Figures 2.10 and 2.11) g (6) has a value of one per cent. In Figure 2.10 x (5) is fixed at 75 per cent. This high x (5)/low g (6) case provides the worst example of a catastrophic collapse of the industry and is clearly unacceptable. Figure 2.11 is the low x (5)/low g (6) case with values of 50 per cent and one per cent respectively. This produces an early peak for the industry but it then remains at a relatively high level of production until 2056 when it collapses in an unacceptable way.

Overall, the case shown here as Figure 2.4, with x (5) fixed at 66.7 per cent and g (6) at three per cent seems to present an acceptable simulation of the real world of oil with respect to: (i) an all over reasonable shape to the discovery and production curves; (ii) discovery of all or most of the resources before the discovery increment approaches zero; and (iii) minimization of the catastrophic fall in the rate of production once the total resource base has been almost or wholly discovered. These values for x (5) and g (6) have, therefore, been used in all the cases presented later in this study.

The parameter m (7) is used to set a certain minimum level of production is some cases where the production growth rates, f (2), are negative. This is used only in the iterations in Chapter 4 in which there is an analysis of the implications of several oil industry studies on the future of oil. In these studies there are uncertainties concerning the length of time over which specific decline rates are supposed to apply. The parameter m (7) is used to eliminate the uncertainty and to set limits on the temporal continuation of decline unconstrained by exogenous variables.

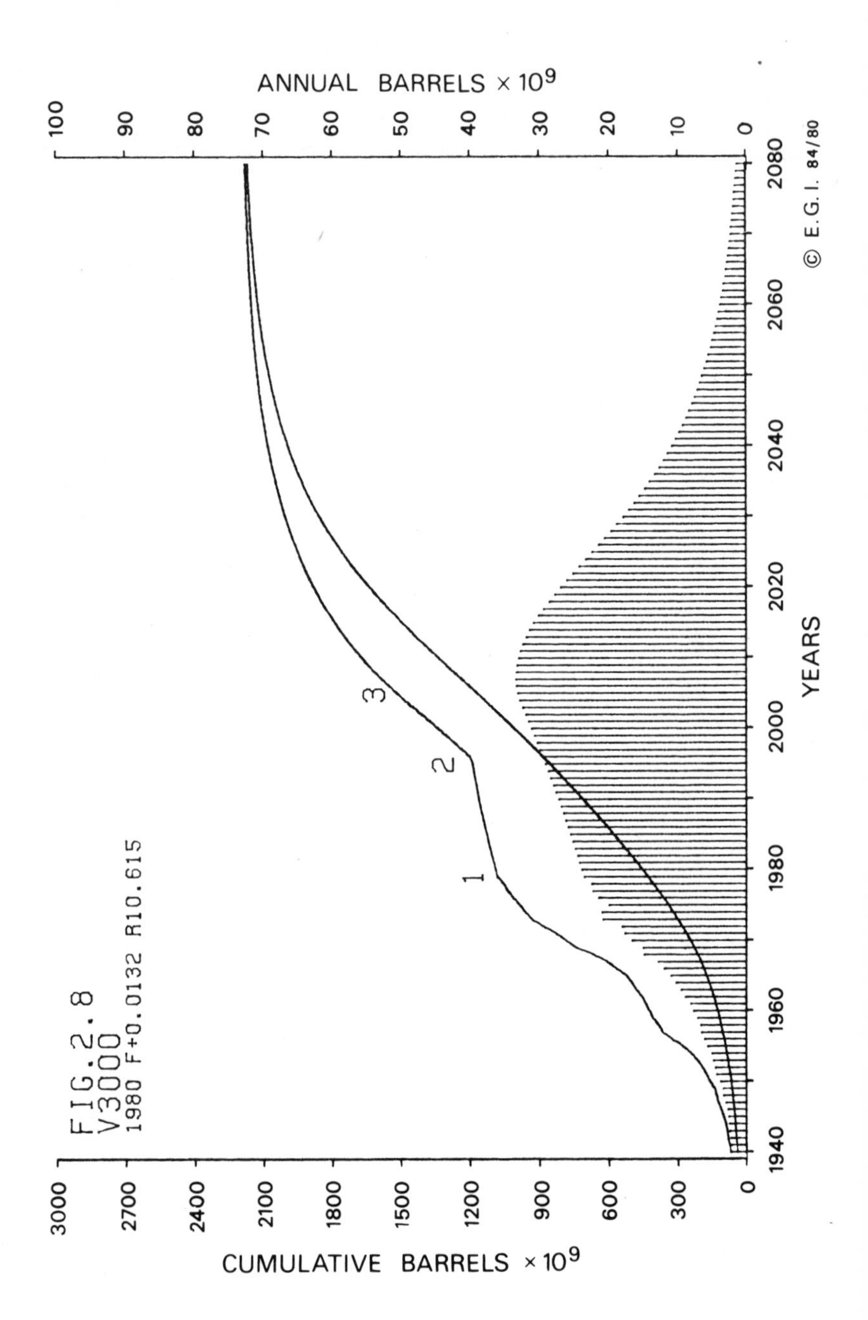
FIG.2.8
V3000
1980 F+0.0132 R10.615
ANNUAL BARRELS × 10^9
100
90
80
70
60
50
40
30
20
10
0
CUMULATIVE BARRELS × 10^9
3000
2700
2400
2100
1800
1500
1200
900
600
300
0
1
2
3
YEARS
1940
1960
1980
2000
2020
2040
2060
2080
© E.G.I. 84/80

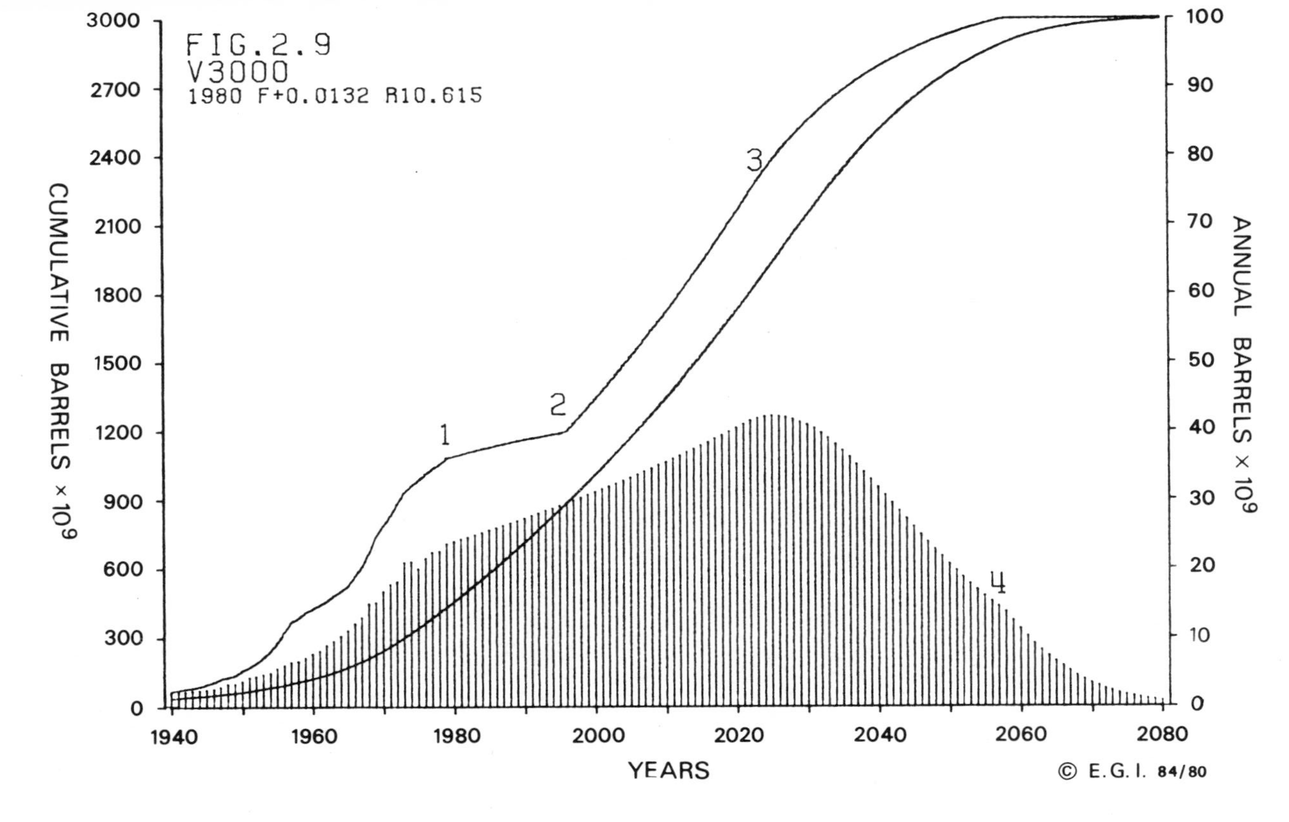

Figure 2.9 *The 3000 x 10^9 Barrels Resource Base, with a 1.32% Growth Rate Case High x (75%), High g (5%)*

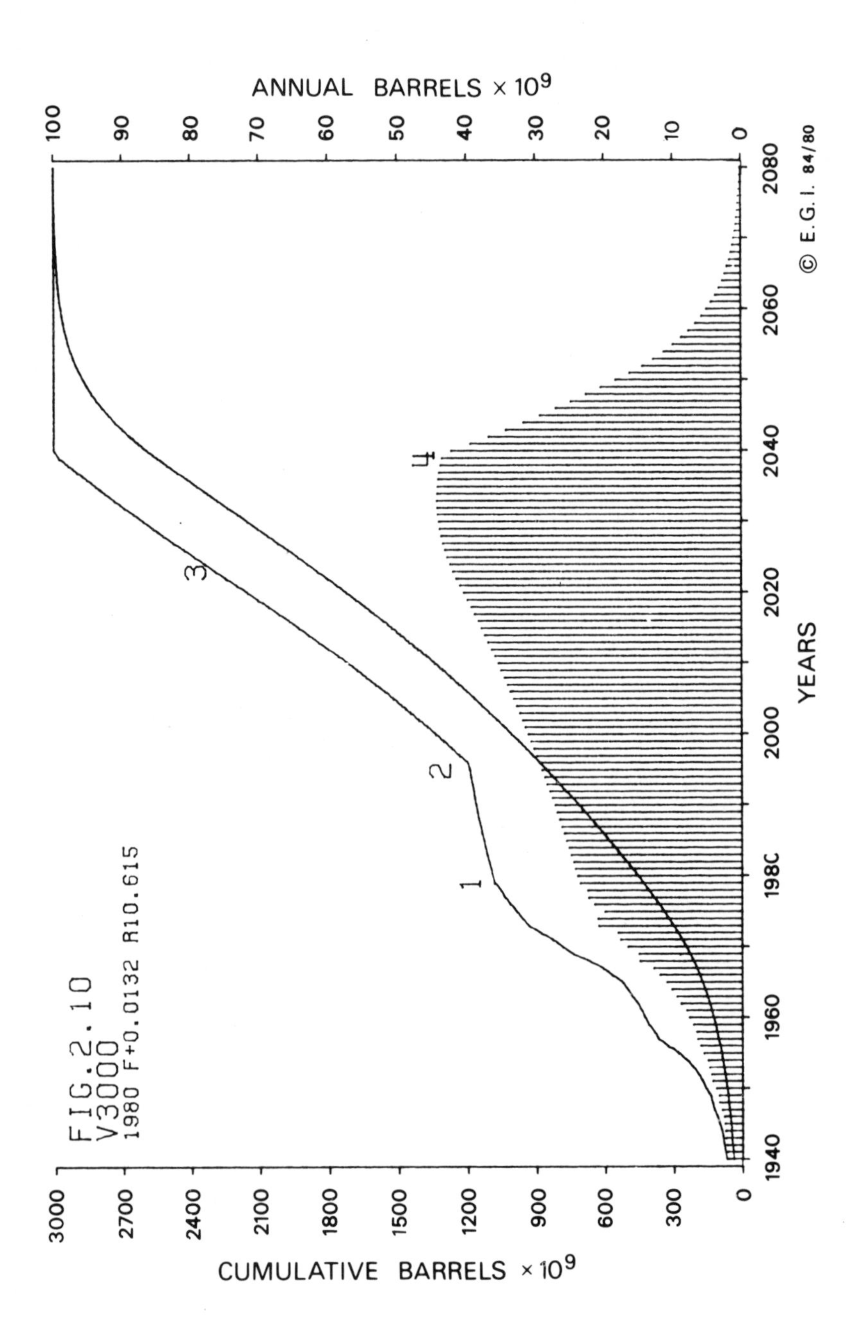
ANNUAL BARRELS × 10⁹
100
90
80
70
60
50
40
30
20
10
0
FIG.2.10
V3000
1980 F+0.0132 R10.615
1
2
3
4
3000
2700
2400
2100
1800
1500
1200
900
600
300
0
CUMULATIVE BARRELS × 10⁹
1940
1960
198C
2000
2020
2040
2060
2080
YEARS
© E.G.I. 84/80

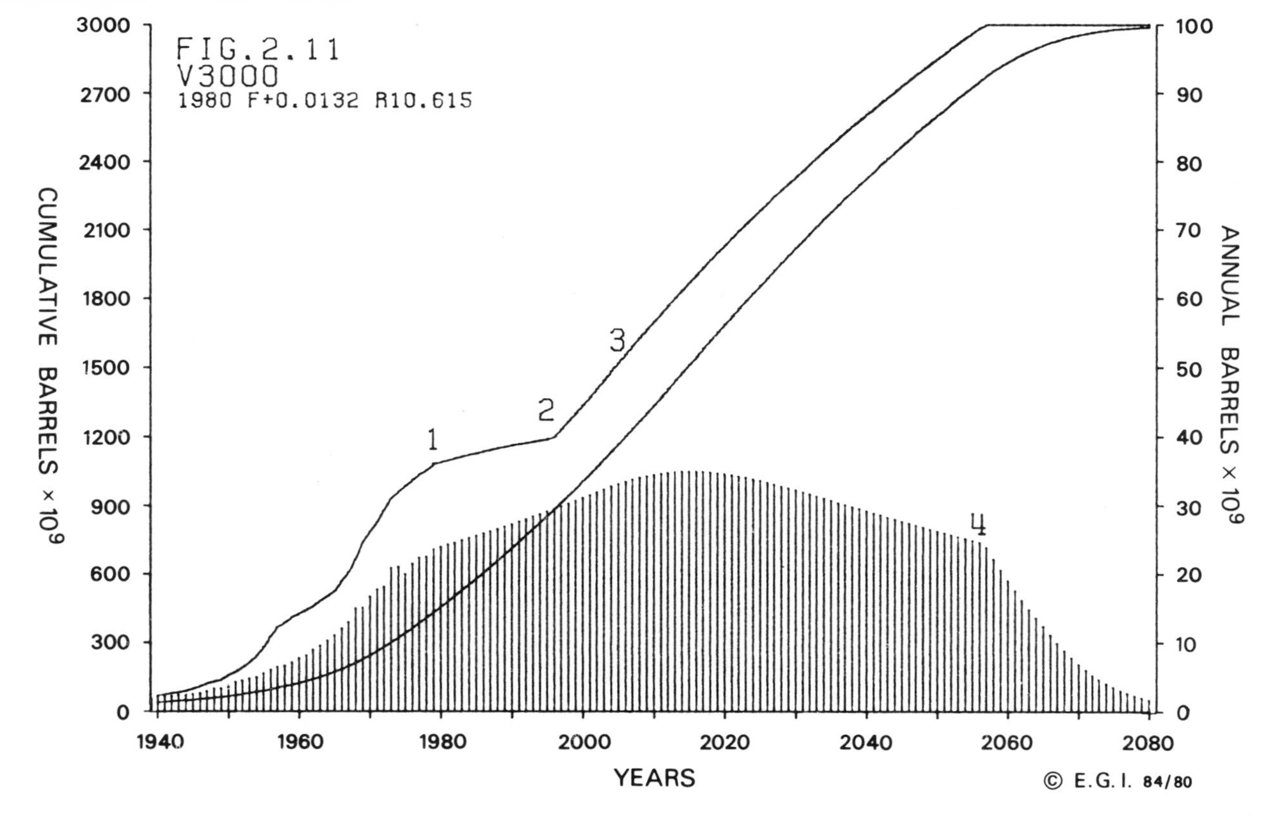

Figure 2.11 *The 3000 x 10^9 Barrels Resource Base, with a 1.32% Growth Rate Case Low x (50%), Low g (1%)*

(ii) *The Output Vectors*

The output of the model consists of four vectors. These are the estimated yearly values for the production of oil, the cumulative production of oil, the annual addition to discovered reserves, and the cumulative discovered reserves. Before defining these vectors, it is necessary to define two additional elements of the notation:

$$t_i = \text{the year, } t_i = 1940, 1941, 1942, \ldots, 2080; \tag{8}$$

$$c = \text{cumulative.} \tag{9}$$

t_i (8) is used as a subscript to denote the yearly values, for each of the four vectors. c (9) is used as a superscript to identify the two vectors of cumulative values. The four vectors of results are then:

$$p_{t_i} = \text{yearly production of oil;} \tag{10}$$

$$s_{t_i} = \text{yearly addition to the discovered reserves;} \tag{11}$$

$$p^c_{ti} = \sum_{k=1}^{i} p_{t_k} = \text{cumulative production of oil to year i;} \tag{12}$$

$$s^c_{ti} = \sum_{k=1}^{i} s_{t_k} = \text{cumulative additions to the discovered reserves to year i.} \tag{13}$$

After running the model on a particular case, each element, t_i, of vector p_{t_i} (10) consists, for the period 1940-79, of the historical volume of production in each year. For the period 1980-2080 each element, t_i, of this vector, p_{t_i} (10), is the estimated production of a particular case. This statement also applies to the vector s^c_{ti} (13) with regard to additions to cumulative known reserves. The vectors p^c_{ti} (12) and s_{t_i} (11) are calculated from p_{t_i} (10) and s^c_{ti} (13) for the period 1940-79 and are estimated for the post-1979 period.

D. The Time Periods

The interrelationships of the parameters and the method of estimating the elements of the four vectors defined above change through time. For this reason the total time span of the model 1940-2080 must be separated into defined time periods. In certain cases, time periods end in specific years while in other cases they end on the satisfaction of some mathematical criteria. For consistency we define the five time periods using set notation. The terms ‘a set of years’ and ‘a time period’ are exactly equivalent.

(i) *Time Period T_1*
This is the historical time period. It begins, arbitrarily, in 1940 and ends in 1979, the last year for which historical data are available:

$$T_1 = \{t_i \mid t_i \geq 1940; t_i \leq 1979\} \text{ for all } i = 1, 2, 3, \ldots, 141. \qquad (14)$$

During this time period the annual level of production, p_{t_i} (10) and the annual cumulative discovered reserves s^c_{ti} (13) are input variables and the annual cumulative production p^c_{ti} (12) is calculated as:

$$p^c_{ti} = \sum_{k=1}^{i} p_{t_k} + 37 \times 10^9 \quad \text{for all } t_i \in T_1; \qquad (15)$$

where 37 x 10^9 is the estimated world oil production before 1940. The annual addition to reserves is simply calculated as:

$$s_{t_i} = s^c_{ti} - s^c_{ti-1} \qquad \text{for all } t_i \in T_1. \qquad (16)$$

Time period T_1 is included in the model simply to complete the overall picture. In all cases the four output vectors are identical.

(ii) *Time Period T_2*
Time period T_2 begins in 1980, the first year for which the four output vectors must be estimated. It is of variable length. At the end of 1979 the actual reserves to production ratio was 27.41 while the value of r (4), as calculated from f (2), range from 10.30 to 14.15 (see Figure 2.2). The parameter r (4) should be used, in this time period, to calculate the annual addition to reserves, s_{t_i} (11), from the annual production, p_{t_i} (10), (see below [27]). If any of the values for r (4), shown in Figure 2.2, were used for this calculation, there would be an immediate large negative addition to discovered reserves, s_{t_i} (11). Since such an 'undiscovery' of reserves is impossible a new vector had to be defined:

$$\hat{r}_{t_i} = (s^c_{ti} - p^c_{ti})/p_{t_i} \qquad \text{for all } t_i \in T_1; \qquad (17)$$

expressing the year end-position of the historical R/P ratio. In time period T_2 then $\hat{r}_{t_i}$ (17) was used to calculate the annual addition to discovered reserves instead of r (4).

$\hat{r}_{t_i}$ (17) has a large value (of 27.41) and it is necessary to reduce this value to the level of r (4) as calculated from the input growth rate, f (2), in order to bring the real-world situation on reserves into line with the needs of the industry based on the required minimum R/P ratio. Logically, the speed of the reduction of $\hat{r}_{t_i}$ (17) during time period T_2 is related to the size of the resource base. Given a resource base of up to 3000×10^9 barrels, the reduction is modelled to be more rapid than with the larger resource bases. This, we would argue, is appropriate because, with the smaller resource bases, the discovery and recovery of additional oil is more difficult, so implying the quicker use of that oil which has already been discovered. Thus, for the smaller resource bases, the value of $\hat{r}_{t_i}$ was reduced each year by two years, whilst for larger resource bases it was reduced by one year each year. Thus, if:

$$v \leq 3000 \times 10^9 \qquad \text{for all } t_i \in T_2; \tag{18}$$

then: $$\hat{r}_{t_i} = \hat{r}_{t_{i-1}} - 2; \tag{19}$$

otherwise: $$\hat{r}_{t_i} = \hat{r}_{t_{i-1}} - 1. \tag{20}$$

In both cases time period T_2 ends when $\hat{r}_{t_i}$ (19 or 20) is less than or equal to the value of r (4). The set T_2 can be defined then as:

$$T_2 = \{t_i | t_i > 1979; \hat{r}_{t_i} > r\} \qquad \text{for all } i = 1, 2, 3, ..., 141 \tag{21}$$

In time period T_2 the critical parameter is f (2), the growth rate in oil production. In some cases the value of f (2) changes over time during the course of the growth period. The years during which this occurs as well as the new growth rates are input vectors of the length four, the maximum number of changes allowed. Each year the current date, t_i (8), is checked against the input list of dates, y_j. If:

$$y_j = t_i \qquad \text{for all } j = 1, 2, 3, 4, \quad \text{for all } t_i \in T_2; \tag{22}$$

then: $$f = \hat{f}_j; \tag{23}$$

and: $$r = 1 + f + f^2 + f^3 + f^4 + f^5 + f^6 + f^7 + f^8 + f^9. \tag{24}$$

Each year's production is calculated from the previous year's production by applying this factor. Thus:

$$p_{t_i} = p_{t_{i-1}} \cdot f \qquad \text{for all } t_t \in T_2. \tag{25}$$

The cumulative production can be calculated simply as the sum of the previous year's cumulative production and this year's production:

$$p^c_{t_i} = p^c_{t_{i-1}} + p_{t_i} \qquad \text{for all } t_i \in T_2. \tag{26}$$

The annual addition to discovered reserves is then calculated as the difference between $\hat{r}_{t_i}$ (19 or 20) multiplied by this year's production, p_{t_i} (10), ie the required remaining discovered reserves and the total reserves discovered up to last year, s^c_{ti-1} (13), less the total production to the end of the present year, p_{t_i} (10). The volume of discovered reserves remaining after this year's production but before this year's discoveries:

$$s_{t_i} = p_{t_i} \cdot \hat{r}_{t_i} - (s^c_{ti-1} - p^c_{ti}) \qquad \text{for all } t_i \in T_2. \qquad (27)$$

Subject to the limitation that s_{t_i} can never be less than zero. Thus if:

$$s_{t_i} < 0.0 \qquad \text{for all } t_i \in T_2; \qquad (28)$$

then: $$s_{t_i} = 0.0. \qquad (29)$$

One of the input parameters is d (3), the maximum amount which can be added to the discovered reserves in any one year. It is possible that the value of s_{t_i} (11) calculated in (27) will exceed d (3). Should this be the case the values calculated in (25), (26) and (27) are all too large and the three values must be recalculated, no longer based on f (2) and $p_{t_{i-1}}$ (25), but rather upon s_{t_i} (11) and d (3). If:

$$s_{t_i} > d \qquad \text{for all } t_i \in T_2; \qquad (30)$$

then: $$s_{t_i} = d. \qquad (31)$$

The production can now be calculated as:

$$p_{t_i} = ((s^c_{ti-1} + s_{t_i}) - p^c_{ti-1})/(1 + \hat{r}_{t_i}). \qquad (32)$$

This value for p_{t_i} (32) will, however, be less than that calculated for p_{t_i} (25), thus violating the requirement that p_{t_i} (25) be equal to $p_{t_{i-1}} \cdot f$. In this case the actual growth rate over the last 10 years is then calculated as:

$$\hat{f}_{t_i} = (p_{t_i}/p_{t_{i-9}})^{1/9}. \qquad (33)$$

The values for p^c_{ti} (12) and s^c_{ti} (13) can be calculated in either case as:

$$p^c_{ti} = p^c_{ti-1} + p_{t_i} \qquad \text{for all } t_i \in T_2; \qquad (34)$$

and: $$s^c_{ti} = s^c_{ti-1} + s_{t_i} \qquad \text{for all } t_i \in T_2. \qquad (35)$$

Thus, in time period T_2, the critical parameter is f (2), the growth rate of production, unless the growth rate causes s_{t_i} (11) to violate the constraint on the growth of reserves d (3), in which case d (3) becomes the critical variable and the growth rate of oil production, f (2), is thus forced to concede.

(iii) *Time Period* T_3

Time period T_3 is also of variable length. It begins when the present excess remaining proven reserves have been reduced so that $\hat{r}_{t_i}$ (19 or 20) is less than or equal to the value of the variable r (24) as calculated for the current f (2). The period ends when more than x (5) per cent of v (1), the total resource base, has been discovered. Set T_3 is then defined as:

$$T_3 = \left\{ t_i \mid \hat{r}_{t_i} \leq r; s^c_{ti} < v \cdot x \right\} \quad \text{for all } i = 1, 2, 3, \ldots, 141. \quad (36)$$

As in time period T_2 the key variable is f (2). However, again as in T_2, it may change in certain years. Therefore, if:

$$y_j = t_i \qquad \text{for all } j = 1, 2, 3, 4, \quad \text{for all } t_i \in T_3; \qquad (37)$$

$$\text{then:} \quad f = \hat{f}_j; \qquad (38)$$

$$\text{and:} \quad r = 1 + f + f^2 + f^3 + f^4 + f^5 + f^6 + f^7 + f^8 + f^9. \qquad (39)$$

The yearly values of each of the four output vectors are calculated in the same manner as that employed in time period T_2, with the exception of (27) and (32) where $\hat{r}_{t_i}$ (17) is replaced by r (4). Note, therefore, that while a new r (24) was calculated for each changed f (2) that value was not used in the calculation of s_{t_i} (27) or p_{t_i} (32) during T_2. Instead $\hat{r}_{t_i}$ (19 or 20) was used. The variable r (24) was used only to identify the end of time period T_2. In time period T_3, however, r (24 or 39) *is* used in these calculations. Thus:

$$p_{t_i} = p_{t_i} \cdot f \qquad \text{for all } t_i \in T_3; \qquad (40)$$

$$p^c_{ti} = p^c_{ti-1} + p_{t_i} \qquad \text{for all } t_i \in T_3; \qquad (41)$$

$$s_{t_i} = p_{t_i} \cdot r - (s^c_{ti-1} - p^c_{ti}) \qquad \text{for all } t_i \in T_3; \qquad (42)$$

$$s^c_{ti} = s^c_{ti-1} + s_{t_i} \qquad \text{for all } t_i \in T_3. \qquad (43)$$

During time period T_3 the growth rate in oil use, f (2), is still the key variable for the calculation of all four output vectors. It is, therefore, possible that the annual additions to discovered reserves, s_{t_i} (42), will be too great. Should that be the case then, just as in time period T_2, the values of p_{t_i} (40), p^c_{ti} (41), s_{t_i} (42) and s^c_{ti} (43) must be recalculated. As in time period T_2, when p_{t_i} (32) and f_{t_i} (33) had to be calculated based on knowledge of s_{t_i} (31), they must be so calculated in this time period as well. This could be done in T_2 because we then knew $\hat{r}_{t_i}$ (19 or 20). In T_3, however, r (39) is a function of f (2) which is in itself unknown when the limitation on s_{t_i} (11) applies. Thus if:

$$s_{t_i} > d \qquad \text{for all } t_i \in T_3; \qquad (44)$$

then:
$$s_{t_i} = d; \qquad (45)$$
$$s^c_{ti} = s^c_{ti-1} + s_{t_i}; \qquad (46)$$
$$\hat{r}_{t_i} = r; \qquad (47)$$
$$p_{t_i} = (s^c_{ti} - p^c_{ti-1})/(1 + \hat{r}_{t_i}); \qquad (48)$$
$$\hat{f}_{t_i} = (p_{t_i}/p_{ti-9})^{1/9}; \qquad (49)$$
$$r' = 1 + \hat{f}_{t_i} + \hat{f}_{t_i}^{\,2} + \hat{f}_{t_i}^{\,3} + \hat{f}_{t_i}^{\,4} + \hat{f}_{t_i}^{\,5} + \hat{f}_{t_i}^{\,6} + \hat{f}_{t_i}^{\,7} + \hat{f}_{t_i}^{\,8} + \hat{f}_{t_i}^{\,9}. \qquad (50)$$

In (47) r, the reserves to production ratio calculated from f in (39 or 24), is used as an estimate of the value of $\hat{r}_{t_i}$. However, since the value of s_{t_i} from (45) is less than the value of s_{t_i} calculated in (42) the (unknown) value of p_{t_i} (48) must also be less than the value of p_{t_i} (40). Therefore, the actual average annual growth rate, $\hat{f}_{t_i}$ (49), over the preceding nine years and this year must be less than the value of f (38 or 23). Likewise, if $\hat{f}_{t_i}$ (49) is less than f (38 or 23), r′ (50) must be less than r (39 or 24). As a result of this overestimate in $\hat{r}_{t_i}$ (47), p_{t_i} (48), $\hat{f}_{t_i}$ (49) and r′ (50) are all underestimates of the correct values. It is possible, however, to find the correct values by iteration. The value r′ (50) is used in (48) instead of $\hat{r}_{t_i}$ (47). p_{t_i} (48), $\hat{f}_{t_i}$ (49) and r′ (50) are recalculated again and again until the difference between any two successive calculations is very small. When this happens the answers have stabilized at a mathematical balance with the value of s_{t_i} (45), to within a small predetermined tolerance. Thus if:

$$|r' - \hat{r}_{t_i}| > 0.001; \qquad (51)$$

then:
$$\hat{r}_{t_i} = r'; \qquad (52)$$

and repeat steps (48), (49), (50), (51); otherwise:

$$\hat{r}_{t_i} = r'. \qquad (53)$$

Finally, it is also possible in certain cases, with negative growth rates, that the level of production may fall below an exogenously defined minimum level. The parameter m (7) can be used to prevent violation of this minimum level of production. If:

$$p_{t_i} < m \qquad \text{for all } t_i \in T_3; \qquad (54)$$

then:
$$p_{t_i} = m; \qquad (55)$$
$$p^c_{ti} = p^c_{ti-1} + p_{t_i}; \qquad (56)$$

$$s_{t_i} = p_{t_i} \cdot r - (s^c_{ti} - p^c_{ti}); \qquad (57)$$

$$s^c_{ti} = s^c_{ti} + s_{t_i}. \qquad (58)$$

Note that it is logically inconsistent, in any one case, for restrictions (44) and (54) to apply at the same time.

(iv) *Time Period T_4*

The preceding time periods have all been periods of growth. With the beginning of time period T_4, we are past the zenith of the oil industry's development and are now concerned with the decline in the volumes of additions to discovered reserves and hence with the decline in production. The period begins when more than x (5) per cent of the total resource base v (1) has been found and ends when all resources have been discovered. The set T_4 can be defined as:

$$T_4 = \left\{ t_i \mid s^c_{ti} \geq x \cdot v;\ s^c_{ti} \geq v \right\} \text{ for all } i = 1, 2, 3, ..., 141. \qquad (59)$$

The total volume of the ultimate resource base, v (1), is a parameter specified exogenously to the model. Logically it constrains both the cumulative additions to discovered reserves, s^c_{ti} (13), and the cumulative production, p^c_{ti} (12), since neither can exceed v (1). The cumulative production p^c_{ti} (12) is also logically limited to be less than or equal to the discovered reserves s^c_{ti} (13) in each year. The critical vectors in this time period are then cumulative discovered reserves, s^c_{ti} (13), and the vector from which it is created, annual additions to discovered reserves, s_{t_i} (12). We have now to concern ourselves with the behaviour of the values of these two through this time period. It is illogical to expect that the volume of additions to known reserves, s_{t_i} (12), will continue to grow, in the manner of the preceding periods, until the last barrel has been proven. Rather the declining volume of the reserves which remain to be proven $(v - s^c_{ti})$ will at some point slow the rate of additions to known reserves. Additions to known reserves, s_{t_i} (12), are hypothesized as coming from two sources; first, from the continuing discovery of new fields; and, second, from the extension of discovered fields and from an improvement of the rate of recovery of oil from each of the fields discovered. Given a finite population of fields of varying sizes and in contrasting locations, one would expect the technically more difficult to detect and/or the smaller fields to be discovered last. Thus over time the rate of return on the exploration effort, in terms of the volume of additional oil for each unit of exploration effort, will be reduced. Second, as the recovery rate is improved each additional improvement in recovery will be technically more difficult and less rewarding in terms of volume per effort unit. The critical parameter thus now becomes g (6) and its effect on the annual volume of additions to discovered reserves s_{t_i} (12). The exact point at which the volume of the resource base, v (1), will become binding on the additions to discovered reserves, s_{t_i} (12), is unpredictable. In his model of oil exploitation for the United States, King Hubbert implicitly assumed that this occurs once 50 per cent of the ultimate resource base has been discovered.[1] In this world model we have assumed 66.7 per cent of total reserves discovered to be a more reasonable figure to use in order to divide the growth period from the decline period.[2] The annual addition to discovered reserves is now calculated as:

$$s_{t_i} = s_{t_{i-1}} \cdot g \qquad \text{for all } t_i \in T_4. \qquad (60)$$

There is the same difficulty with the calculation of annual production p_{t_i} (10), from annual additions to discovered reserves, s_{t_i} (11), as was noted in period T_3. The problem is again solved by iteratively calculating p_{t_i} (63), $\hat{f}_{t_i}$ (64) and $\hat{r}_{t_i}$ (65 and 67). This procedure is more fully discussed in time period T_3 above (see p. 73). Here we note only that the procedure converges. Therefore:

$$s^c_{t_i} = s^c_{t_{i-1}} + s_{t_i} \qquad \text{for all } t_i \in T_4; \qquad (61)$$

$$\hat{r}_{t_i} = \hat{r}_{t_{i-1}} \qquad \text{for all } t_i \in T_4; \qquad (62)$$

$$p_{t_i} = (s^c_{t_i} - p^c_{t_{i-1}})/(1 + \hat{r}_{t_i}) \qquad \text{for all } t_i \in T_4; \qquad (63)$$

$$\hat{f}_{t_i} = (p_{t_i}/p_{t_{i-9}})^{1/9} \qquad \text{for all } t_i \in T_4; \qquad (64)$$

$$r' = 1 + \hat{f}_{t_i} + \hat{f}_{t_i}^{\,2} + \hat{f}_{t_i}^{\,3} + \hat{f}_{t_i}^{\,4} + \hat{f}_{t_i}^{\,5} + \hat{f}_{t_i}^{\,6} + \hat{f}_{t_i}^{\,7} + \hat{f}_{t_i}^{\,8} + \hat{f}_{t_i}^{\,9}; \qquad (65)$$

$$\text{and if: } |r' - \hat{r}_{t_i}| > 0.001 \qquad \text{for all } t_i \in T_4; \qquad (66)$$

1. M.King Hubbert, 'Hubbert estimates from 1956 to 1974 of US Oil and Gas' in: M.Grenon (Ed.), *Methods and Models for Assessing Energy Resources*, Oxford, Pergamon Press, 1979, pp.370-383, particularly Figures 1, 3, 4, 5 and 7. The shape of each of these curves is dependent upon Hubbert's implicit assumption that the 50 per cent of the ultimate resource base being discovered will divide growth from decline, thus yielding his symmetrical diagrams. At the IIASA/RSI Conference at Laxenburg, 9-14 July 1979, Hubbert stated in discussion that the symmetrical form of the curves (which is common to all statements of his work) was not necessary. He refused, however, to speculate on an alternative form. See, M.Grenon, *Systems Aspects of Energy and Mineral Resources*, Pergamon Press, Oxford, (forthcoming).
2. For the justification of this choice, see above pp. 54 to 56 and Figures 2.2 to 2.11.

then:	$\hat{r}_{t_i} = r'$		(67)
and:	repeat steps (63), (64), (65) and (66);		
otherwise:	$\hat{r}_{t_i} = r'$	for all $t_i \in T_4$;	(68)
and:	$p^c_{t_i} = p^c_{t_{i-1}} + p_{t_i}$	for all $t_i \in T_4$.	(69)

These procedures are used until the end of time period T_4 which occurs when the resource base is all discovered:

$$s^c_{t_i} \geq v \qquad \text{for all } t_i \in T_4. \qquad (70)$$

(v) *Time Period T_5*
Time period T_5 is basically the tail-off period of the industry. It is the period which is furthest in the future and as such is the least important. It is of variable length since it begins when all recoverable reserves are discovered. In the model, however, the period always ends in the year 2080 whether or not all oil is discovered and used. The set T_5 can be defined as:

$$T_5 = \left\{t_i \mid s^c_{t_i} \geq v\right\} \qquad \text{for all } i = 1, 2, 3, ..., 141. \qquad (71)$$

The cumulative additions to discovered reserves, $s^c_{t_i}$ (13), is always set equal to the parameter v (1):

$$s^c_{t_i} = v \qquad \text{for all } t_i \in T_5; \qquad (72)$$

and the additions to reserves, s_{t_i} (11), are calculated by subtraction:

$$s_{t_i} = s^c_{t_i} - s^c_{t_{i-1}} \qquad \text{for all } t_i \in T_5. \qquad (73)$$

The cumulative discovered reserves, $s^c_{t_i}$ (13), are thus a constant in period T_5 as are the additions to the discovered reserves, s_{t_i} (11), after the first year. The annual value of oil production, p_{t_i} (10), is calculated as $1/\hat{r}_{t_i}$ of the remaining unproduced reserves. However, as in time periods T_3 and T_4, the reserves to production ratio, $\hat{r}_{t_i}$ (50 and 52), is a function of the growth rate, f (2), and the growth rate is a function of the production, p_{t_i} (10). The iterative procedure described in time period T_3 (48), (49), (50) and (51) was again used here:

$$p_{t_i} = (s^c_{t_i} - p^c_{t_{i-1}})/(1 + \hat{r}_{t_i}) \qquad \text{for all } t_i \in T_5; \qquad (74)$$

$$\hat{f}_{t_i} = (p_{t_i}/p_{t_{i-9}})^{1/9} \qquad \text{for all } t_i \in T_5; \qquad (75)$$

$$r' = 1 + \hat{f}_{t_i} + \hat{f}_{t_i}^{\,2} + \hat{f}_{t_i}^{\,3} + \hat{f}_{t_i}^{\,4} + \hat{f}_{t_i}^{\,5} + \hat{f}_{t_i}^{\,6} + \hat{f}_{t_i}^{\,7} + \hat{f}_{t_i}^{\,8} + \hat{f}_{t_i}^{\,9}; \qquad (76)$$

if:	$\lvert r' - \hat{r}_{t_i} \rvert > 0.001$	for all $t_i \in T_5$;	(77)
then:	$\hat{r}_{t_i} = r'$		(78)
and:	repeat steps (73), (74), (75) and (76);		
otherwise:	$\hat{r}_{t_i} = r'$	for all $t_i \in T_5$.	(79)

Finally the cumulative production is calculated as:

$$p^c_{ti} = p^c_{ti-1} + p_{ti} \qquad \text{for all } t_i \in T_5. \qquad (80)$$

The time period, and the model, ends when the values for the year 2080 have been calculated.

E. The Incidence of the Time Periods in Iterations of the Model

Time periods T_1 (14) and T_2 (21) must occur in all of the iterations. Time period T_3 (36) need not always appear; however, with the parameters we have used in our set of cases it does, in fact, always occur. The other time periods T_4 (59) and/or T_5 (71) may or may not appear in a specific case. With large resource bases and low growth rates time periods T_4 (59) and T_5 (71), or T_5 (71) alone, may only occur post the year 2080 and thus they will not appear in the model which is limited to the period up to 2080.

F. Reading the Graphs

The model was coded in Fortran IV and was run on the IBM 370/158 at the Technical University of Delft. Two sorts of output were created. First, there is a printed list of the estimates of the annual values of each of the four output vectors interspersed with messages to identify which set of equations were being used. Important values from these lists are presented in the various tables. Second, there is graphical output. This forms the basis of the illustrations presented in this study. Figure 2.12 is an example of one of these diagrams.

The information given in the upper left-hand corner of Figure 2.12, and in each of the other graphs presenting an iteration of the model, identifies the figure number and also gives the values of the relevant parameters for that case. The way in which this information is presented is shown in Figure 2.13.

The first line of the text gives the figure-number. The next line begins with a 'V' and the number which follows this specifies

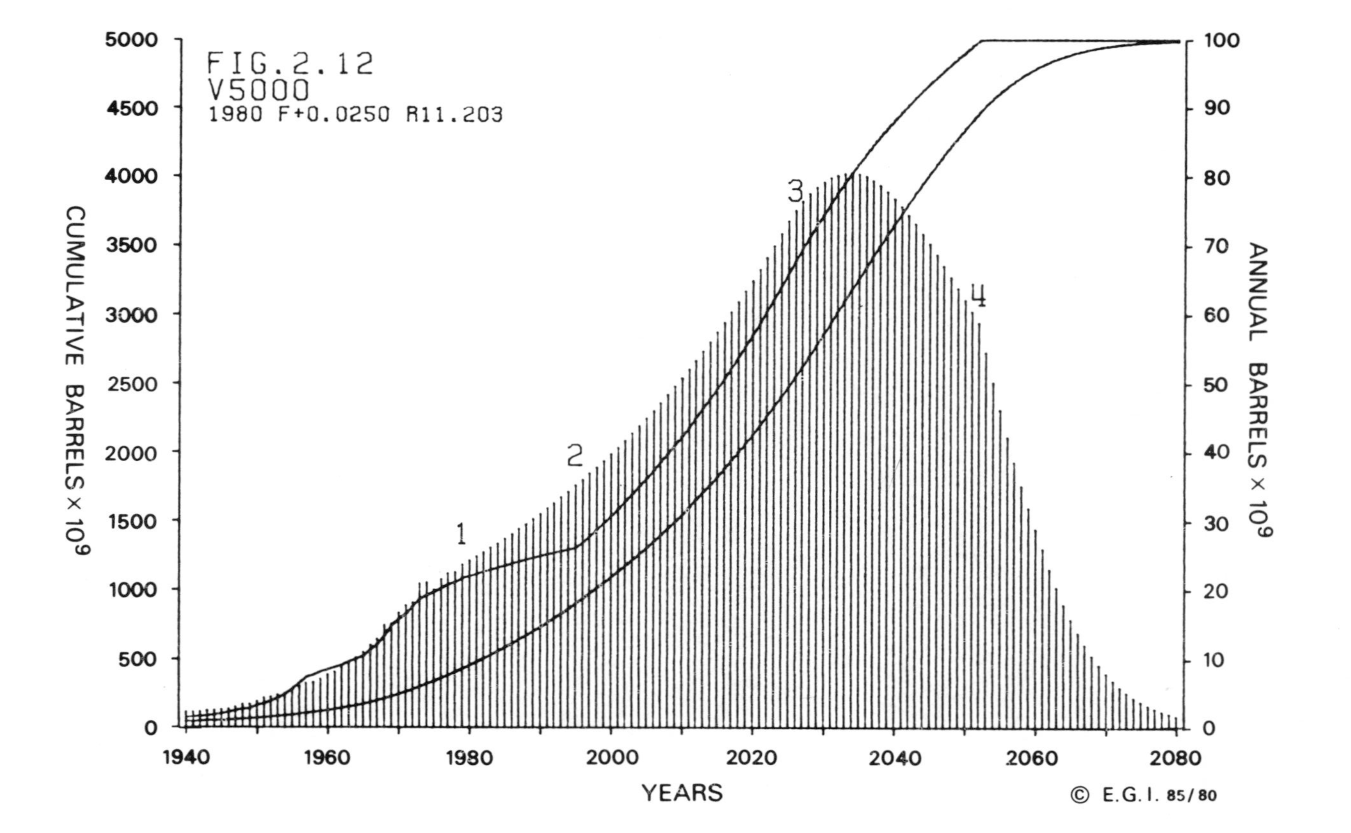

Figure 2.12 *The 5000 x 10^9 Barrels Resource Base, with a 2.5% Growth Rate Case*

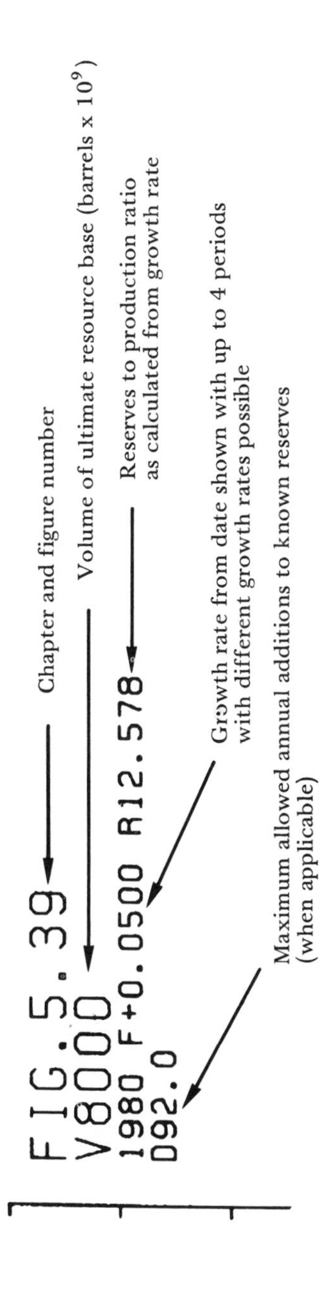

Figure 2.13 *The Values of the Input Parameters: an Annotated Diagram*

the size of the ultimate resource base (in barrels x 10^9) with which the case is working. The next one, two, three or four (maximum) lines give information on the growth rates in oil use in time periods T_2 and T_3. Each line (or the single line) defines the growth rate used for a number of years in the time periods (or for the whole period). Each line begins with the year in which the growth rate first applies. This is followed by the letter 'F' preceding the applicable positive or negative growth rate. There then follows the letter 'R' and the calculated effective R/P ratio for that growth rate. In Figure 2.12 only one growth rate is used so there is only one line of information.

The last line of information on the diagram may begin with the letter 'D'. When this occurs it is followed by the maximum allowed volume of annual additions to known reserves. Note, however, that this line only appears in those cases in which there is a limitation to reserves additions. Alternatively the last line can, in a few cases, begin with an 'M'. This occurs when there are cases with negative growth rates in time periods T_2 and T_3 and indicates the minimum allowed production in that phase of the iteration.

Three of the four output vectors are graphically portrayed on these diagrams; *viz* cumulative discoveries, $s^c_{t_i}$ (13); cumulative production, $p^c_{t_i}$ (12); and annual production, p_{t_i} (10). The first two are read from the left-hand scale and the third, p_{t_i} (10), is read from the right-hand vertical scale. This latter scale is of constant length in all presentations. It runs from 0.0 to 100 x 10^9 barrels. Each production curve, displayed as a series of vertical lines (one for each year), is, therefore, on the same scale and thus directly comparable with all others, except when it has been necessary, for presentation purposes, to reduce the scale of an illustration to fit the page size. The small number that fall into this category are printed vertically on the page (eg Figure 5.25 on p. 227) rather than across it.

This is not the case in respect of the two cumulative vectors which are displayed as lines curving from the lower left corner up to the upper right corner of the diagram. These lines, read from the left-hand vertical scale, have had to be scaled according to the size of the resource base, v (1), for reasons of presentation in published form. In each case the left-hand scale runs from 0.0 to v (1). The use of different scales is related to the very wide range of values for v (1). This sort of presentation means that visual comparisons of the shapes of the cumulative discovery curve, $s^c_{t_i}$ (13), and cumulative production curve, $p^c_{t_i}$ (12) are valid only between cases which have the same value of v (1).

The top line is the curve of cumulative additions to discovered reserves, $s^c_{t_i}$ (13). The lower curve is the cumulative production, $p^c_{t_i}$ (12). Since these curves relate to a scale which is different from that of the production curve, p_{t_i} (10), there can never be direct comparison between them.

Finally, the numbers 1 and 2 appear on every diagram and the numbers 3 and 4 on most of them. These numbers are located above the x axis to show the last year of the time periods 1, 2, 3 and 4, respectively.

Appendix 2.1

Listing of the Computer Program

The following computer program was used to run all the cases presented in this book.

Comment cards have been inserted to identify the time periods as discussed in this chapter.

A section of program to plot the results follows the end of time period T_5. This contains a series of call statements to the CALCOMP routines used to do the plotting work. Computer installations which do not have these facilities available will have to replace these call statements with appropriate instructions for their own plotting software, if available. Otherwise that section of program should be discarded. Removal of the section will not affect the numerical results but it will, of course, mean that no graphic output will be generated.

```
      DIMENSION CUMRES(200),CUMPRO(200),PROD(200),X(200),
     . RESINC(200),IBCD(2),JAAR(4),FACT(4),RPP(4)
      CHCK=0.0
      READ(5,10) IBCD,PLAT,FLIM,TV,PROP,DP,
     .  (JAAR(I),FACT(I),I=1,4)
   10 FORMAT(2A4,2X,2F10.0/3F10.0/4(I5,F10.0))
      PLAT IS THE PLATEAU LEVEL FOR NEGATIVE RATES
      FLIM IS THE MAXIMUM ON ADDITIONS TO RESEPVES
      TV IS TOTAL VOLUME OF POSSIBLE RESERVES
      FAC IS GROWTH RATE OF PRODUCTION
      PROP IS PROPORTION OF TV FOUND BEFORE TAIL OFF
      DP IS THE RATE OF TAIL OFF
      WRITE(6,1000) TV,(JAAR(I),FACT(I),I=1,4),PROP,DP,IBCD
 1000 FORMAT(' THE TOTAL VOLUME IS ',F10.1//
     . ' THE PRODUCTION GROWTH RATE ',I4,' IS ',F10.4/
     . ' AFTER ',I4,' CHANGED TO ',F10.4/
     . ' AFTER ',I4,' CHANGED TO ',F10.4/
     . ' AFTER ',I4,' CHANGED TO ',F10.4//
     . ' THE PROPORTION OF THE TOTAL VOLUME FOUND BEFORE ',
     . 'TAIL OFF BEGINS IS ',F10.5//
     . ' THE TAIL-OFF OF DISCOVERY IS AT ',F10.4//////
     . 50X,' THIS IS CASE NUMBER ',2A4/////)
      FAC=FACT(1)
      RP=1+FAC+FAC**2+FAC**3+FAC**4+FAC**5+FAC**6+FAC**7+FAC**8+FAC**9
      RPP(1)=RP
```

```
      L77=0
      IHOLD=0
      READ(5,2) PROD(1),CUMRES(1)
    2 FORMAT(2F10.0)
      CUMPRO(1)=PROD(1)+37.0
      WRITE(6,30) PROD(1),CUMPRO(1),CUMRES(1)
   30 FORMAT('  READING, PRODUCTION,CUM.PROD.,CUM.RESERVES'
     . ,',RESERVES INC.',/'  1940',4X,F10.3,F11.3,F12.3)
      II=2
    1 READ(5,2,END=1111) PROD(II),CUMRES(II)
      I=II-1
      CUMPRO(II)=CUMPRO(I)+PROD(II)
      RESINC(II)=CUMRES(II)-CUMRES(I)
      KY=1939+II
      WRITE(6,20) KY,PROD(II),CUMPRO(II),CUMRES(II),RESINC(II)
   20 FORMAT(I6,4X,F10.3,F11.3,F12.3,F13.3)
      II=II+1
      GO TO 1
 1111 III=II
      II=II-1
      J=II
      RPT=(CUMRES(J)-CUMPRO(J))/PROD(J)
      HTV=TV*PROP
      WRITE(6,31) FAC,RP
   31 FORMAT(/'  PRODUCTION INCREASING AT ',F10.4,' RP IS ',F12.7)
    3 PROD(III)=PROD(J)*FAC
      IIII=III+1939
      DO 1010 I=2,4
      IF(IIII .NE. JAAR(I)) GO TO 1010
      FAC=FACT(I)
      RP=1+FAC+FAC**2+FAC**3+FAC**4+FAC**5+FAC**6+FAC**7+FAC**8+FAC*
      RPP(I)=RP
      WRITE(6,31) FAC,RP
 1010 CONTINUE
      IF(PROD(III) .GE. PLAT) GO TO 1001
      IF(FAC .GT. 1.0) GO TO 1001
      RP=10.0
      PROD(III)=PLAT
      FAC=1.0
      WRITE(6,1002) FAC
 1002 FORMAT(/' GROWTH RATE RESET TO ',F10.5)
 1001 CONTINUE
      CUMPRO(III)=CUMPRO(J)+PROD(III)
      CUMRES(III)=PROD(III)*RPT+CUMPRO(J)
      IF(CUMRES(III) .LT. CUMRES(J)) CUMRES(III)=CUMRES(J)
      RESINC(III)=CUMRES(III)-CUMRES(J)
      IF(RESINC(III) .LT. FLIM) GO TO 6004
      WRITE(6,6001) RESINC(III),FLIM
      RESINC(III)=FLIM
      CUMRES(III) =CUMRES(J)+RESINC(III)
      IF(CHCK .GT. 1.0) WRITE(6,3838) CUMRES(III),CUMPRO(J),RPT
      PROD(III)=(CUMRES(III)-CUMPRO(J))/(1+RPT)
      FAC=(PROD(III)/PROD(III-9))**(1.0/9.0)
      RP1=1+FAC+FAC**2+FAC**3+FAC**4+FAC**5+FAC**6+FAC**7+FAC**8+FAC
      IF(CHCK .GT. 1.0) WRITE(6,3839) FAC,RP1
      KY=III+1939
      WRITE(6,6005) KY,RP1,FA
 6005 FORMAT(I6,' ACTUAL R/P RATIO THIS YEAR IS ',F12.8,
     . ' ACTUAL GROWTH RATE IS ',F10.5)
 6004 CONTINUE
      RPT=RPT-1
      IF (TV .LT. 3001) RPT=RPT-1
      CUMPRO(III)=PROD(III)+CUMPRO(J)
      IF(RPT .LT.RP) GO TO 4
```

```
      CUMRES(III)=CUMRES(J)+RESINC(III)
      KY=III+1939
      WRITE(6,23) KY,PROD(III),CUMPRO(III),CUMRES(III),RESINC(III),RPT
      III=III+1
      IF(III .GT. 141) GO TO 100
      J=J+1
      GO TO 3
    4 CONTINUE
      JJ=J
      WRITE(6,32)
   32 FORMAT(/'  CUMULATIVE RESERVES NOW CALCULATED FROM PRODUCTION')
    5 CONTINUE
      PROD(III)= PROD(JJ)*FAC
      IIII=III+1939
      DO 1012 I=2,4
      IF(IIII .NE. JAAR(I)) GO TO 1012
      FAC=FACT(I)
      RP=1+FAC+FAC**2+FAC**3+FAC**4+FAC**5+FAC**6+FAC**7+FAC**8+FAC**9
      RPP(I)=RP
      WRITE(6,31) FAC,RP
   12 CONTINUE
      IF(PROD(III) .GE. PLAT) GO TO 1005
      IF(FAC .GT. 1.0) GO TO 1005
      PROD(III)=PLAT
      FAC=1.0
      RP=10.0
      WRITE(6,1002) FAC
   05 CUMPRO(III)=CUMPRO(JJ)+PROD(III)
      K=JJ
      CUMRES(III)=CUMPRO(III)+PROD(III)*RP
      IF(CUMRES(III) .LT. CUMRES(JJ)) CUMRES(III)=CUMRES(JJ)
      RESINC(III)=CUMRES(III)-CUMRES(JJ)
      IF(RESINC(III) .LE. FLIM) GO TO 6000
      WRITE(6,6001) RESINC(III),FLIM
   01 FORMAT(/'  RESERVES INCREMENT OF ',F8.2,' TO GREAT,'/
     . '  RESET TO ',F8.2)
      RESINC(III)=FLIM
      CUMRES(III)=CUMRES(JJ)+RESINC(III)
   41 IF(CHCK .GT. 1.0) WRITE(6,3838) CUMRES(III),CUMPRO(JJ),RP
      PROD(III)=(CUMRES(III)-CUMPRO(JJ))/(1+RP)
      CUMPRO(III)=CUMPRO(JJ)+PROD(III)
      FAC=(PROD(III)/PROD(III-9))**(1.0/9.0)
      RP1=1+FAC+FAC**2+FAC**3+FAC**4+FAC**5+FAC**6+FAC**7+FAC**8+FAC**9
      IF(RP1 .LT. 1.0) RP1=1.0
      IF(CHCK .GT. 1.0) WRITE(6,3839) FAC,RP1
      RP2=ABS(RP-RP1)
      IF(RP2 .LT. 0.001) GO TO 3740
      RP=RP1
      GO TO 3741
   40 CONTINUE
      KY=III+1939
      WRITE(6,6005) KY,RP1,FAC
   00 CONTINUE
      IF(CUMRES(III) .GT. HTV) GO TO 12
      KY=III+1939
      WRITE(6,20) KY,PROD(III),CUMPRO(III),CUMRES(III),RESINC(III)
      III=III+1
      JJ=JJ+1
      IF(III .GT. 141) GO TO 100
      GO TO 5
   12 CONTINUE
      IHOLD=(III)
      WRITE(6,33) HTV,PROP,DP
```

```
   33 FORMAT(/'  CUM. RESERVES OVER ',F10.3,' WHICH IS ',
     .  F10.3,' OF TOTAL'/
     .  '  GROWTH OF RESERVES DECREASING AT RATE OF ',F10.3)
      L=JJ-1
    6 CUMRES(III)=(CUMRES(JJ)-CUMRES(L))*(1.0-DP)
      CUMRES(III)=CUMRES(JJ)+CUMRES(III)
      RESINC(III)=CUMRES(III)-CUMRES(JJ)
      IF(CUMRES(III) .GE. TV) GO TO 13
 3841 IF(CHCK .GT. 1.0) WRITE(6,3838) CUMRES(III),CUMPRO(JJ),RP
 3838 FORMAT(1X,'CUMRES= ',F14.5,'OLD CUMPRO= ',F14.5,' RP TO CALC. F
     . ,'= ',F14.5)
      PROD(III)=(CUMRES(III)-CUMPRO(JJ))/(1+RP)
      FAC=(PROD(III)/PROD(III-9))**(1.0/9.0)
      RP1=1+FAC+FAC**2+FAC**3+FAC**4+FAC**5+FAC**6+FAC**7+FAC**8+FAC*
      IF(RP1 .LT. 1.0) RP1=1.0
      IF(CHCK .GT. 1.0) WRITE(6,3839) FAC,RP1
 3839 FORMAT(' FAC FROM OLD RP= ',F14.5,' NEW RP= ',F14.5/)
      RP2=ABS(RP-RP1)
      IF (RP2 .LT. 0.001) GO TO 3840
      RP=RP1
      GO TO 3841
 3840 CONTINUE
      CUMPRO(III)=CUMPRO(JJ)+PROD(III)
      IF(PROD(III) .GT. PROD(K)) K=III
      KY=III+1939
      WRITE(6,23)KY,PROD(III),CUMPRO(III),CUMRES(III),RESINC(III),RP,
   23 FORMAT(I6,4X,F10.3,F11.3,F12.3,F13.3,' RES.-PROD. RATIO = ',F10
     .    ' GROWTH RATE IS ',F10.5//)
      III=III+1
      JJ=JJ+1
      L=L+1
      IF(III .GT. 141) GO TO 100
      GO TO 6
   13 CONTINUE
      WRITE(6,35)
      L77=L
   35 FORMAT('  CUMULATIVE RESERVES AT MAXIMUM ')
    7 CUMRES(III)=TV
      RESINC(III)=CUMRES(III)-CUMRES(JJ)
 3941 IF(CHCK .GT. 1.0) WRITE(6,3838) CUMRES(III),CUMPRO(JJ),RP
      PROD(III)=(CUMRES(III)-CUMPRO(JJ))/(1+RP)
      FAC=(PROD(III)/PROD(III-9))**(1.0/9.0)
      RP1=1+FAC+FAC**2+FAC**3+FAC**4+FAC**5+FAC**6+FAC**7+FAC**8+FAC*
      IF(RP1 .LT. 1.0) RP1=1.0
      IF(CHCK .GT. 1.0) WRITE(6,3839) FAC,RP1
      RP2=ABS(RP-RP1)
      IF(RP2 .LT. 0.001) GO TO 3940
      RP=RP1
      GO TO 3941
 3940 CONTINUE
      CUMPRO(III)=CUMPRO(JJ)+PROD(III)
      IF(CUMPRO(III) .GE. TV) GO TO 100
      KY=III+1939
      WRITE(6,23)KY,PROD(III),CUMPRO(III),CUMRES(III),RESINC(III),RP,
      RP=RP1
      III=III+1
      IF(III .GT. 141) GO TO 100
      JJ=JJ+1
      GO TO 7
   11 CUMPRO(III)=TV
      PROD(III)=CUMPRO(III)-CUMPRO(JJ)
      CUMRES(III)=TV
      RESINC(III)=CUMRES(III)-CUMRES(JJ)
```

```
100 III=III-1
    KK=K+1939
    FD=III
    DO 101 I=1,III
    F=I
    X(I)=F/10.0+0.5
    FDD=FD/10.0
    TVV=TV/10.0
    TVV=TVV/100.0
    CUMPRO(I)=CUMPRO(I)/TV*10.0+0.5
    CUMRES(I)=CUMRES(I)/TV*10.0+0.5
    RESINC(I)=RESINC(I)/10.0+0.5
101 PROD(I)=PROD(I)/10.0+0.5
    IHOLD=IHOLD-1
    CALL PLOTS(1,20)
    CALL FACTOR(0.37)
    CALL PLOT(0.5,0.5,3)
    CALL PLOT(0.4,0.5,2)
    AX=0.5
    DO 4743 I=1,10
    CALL PLOT(0.5,AX,2)
    AX=AX+1.0
    CALL PLOT(0.5,AX,2)
4743 CALL PLOT(0.4,AX,2)
    CALL PLOT(0.5,AX,2)
    CALL PLOT(0.5,0.5,2)
    CALL PLOT(0.6,0.5,3)
    CALL PLOT(0.6,0.4,2)
    AX=0.6
    DO 4744 I=1,14
    CALL PLOT(AX,0.5,2)
    AX=AX+1.0
    CALL PLOT(AX,0.5,2)
4744 CALL PLOT(AX,0.4,2)
    CALL PLOT(14.7,0.5,3)
    CALL PLOT(14.8,0.5,2)
    AX=0.5
    DO 4745 I=1,10
    CALL PLOT(14.7,AX,2)
    AX=AX+1.0
    CALL PLOT(14.7,AX,2)
4745 CALL PLOT(14.8,AX,2)
    CALL PLOT(14.7,AX,2)
    CALL PLOT(14.7,0.5,2)
    CALL PLOT(14.6,0.5,3)
    CALL PLOT(0.6,0.5,2)
    DO 8800 LLKKK=1,2
    CALL SYMBOL(0.9,10.0,0.3,IBCD,0.0,8)
    CALL SYMBOL(0.9,9.6,0.3,101,0.0,-1)
    CALL NUMBER(999.0,9.6,0.3,TV,0.0,-1)
    B=9.3
    DO 4000 I=1,4
    IF(JAAR(I) .LT. 1) GO TO 4000
    FF=JAAR(I)
    CALL NUMBER(0.9,B,0.2,FF,0.0,-1)
    CALL SYMBOL(999.0,B,0.2,64,0.0,-1)
    CALL SYMBOL(999.0,B,0.2,70,0.0,-1)
    FF=FACT(I)-1.0
    IF(FF .GE. 0.0) CALL SYMBOL(999.0,B,0.2,78,0.0,-1)
    CALL NUMBER(999.0,B,0.2,FF,0.0,4)
    CALL SYMBOL(999.0,B,0.2,64,0.0,-1)
    CALL SYMBOL(999.0,B,0.2,89,0.0,-1)
    FF=RPP(I)
```

```
      CALL NUMBER(999.0,B,0.2,FF,0.0,3)
      B=B-0.3
4000  CONTINUE
      IF(PLAT .LE. 0.0001) GO TO 4741
      CALL SYMBOL(0.9,B,0.2,84,0.0,-1)
      CALL NUMBER(999.0,B,0.2,PLAT,0.0,1)
4741  CONTINUE
      IF(FLIM .GE. 900.0) GO TO 4742
      CALL SYMBOL(0.9,B,0.2,68,0.0,-1)
      CALL NUMBER(999.0,B,0.2,FLIM,0.0,1)
4742  CONTINUE
8800  CONTINUE
      CALL PLOT(X(1),CUMRES(1),3)
      DO 102 I=1,II
      KX1=I-1
 102  CALL PLOT(X(I),CUMRES(I),2)
      X1=X(KX1)
      Y1=AMAX1(CUMRES(KX1),PROD(KX1+1))
      Y1=Y1+0.3
      CALL SYMBOL(X1,Y1,0.3,113,0.0,-1)
      CALL SYMBOL(X1,Y1,0.3,113,0.0,-1)
      CALL PLOT(X(II),CUMRES(II),3)
      DO 103 I=II,J
      KX1=I-1
 103  CALL PLOT(X(I),CUMRES(I),2)
      X1=X(KX1)
      Y1=AMAX1(CUMRES(KX1),PROD(KX1+1))
      Y1=Y1+0.3
      CALL SYMBOL(X1,Y1,0.3,114,0.0,-1)
      CALL SYMBOL(X1,Y1,0.3,114,0.0,-1)
      CALL PLOT(X(J),CUMRES(J),3)
      DO 104 I=J,III
 104  CALL PLOT(X(I),CUMRES(I),2)
      IF(L77 .EQ. 0) GO TO 7000
      X1=X(L77)
      Y1=PROD(L77)+0.1
      CALL SYMBOL(X1,Y1,0.3,116,0.0,-1)
      CALL SYMBOL(X1,Y1,0.3,116,0.0,-1)
7000  CONTINUE
      IF (IHOLD .LE. 1) GO TO 7001
      X1=X(IHOLD)
      Y1=AMAX1(CUMRES(IHOLD),PROD(IHOLD))
      Y1=Y1+0.3
      CALL SYMBOL(X1,Y1,0.3,115,0.0,-1)
      CALL SYMBOL(X1,Y1,0.3,115,0.0,-1)
7001  CONTINUE
      CALL PLOT(X(1),CUMPRO(1),3)
      DO 105 I=1,III
 105  CALL PLOT(X(I),CUMPRO(I),2)
      I2=III-1
      DO 106 I=1,I2,2
      I1=I+1
      CALL PLOT(X(I),0.5,3)
      CALL PLOT(X(I),PROD(I),2)
      CALL PLOT(X(I1),PROD(I1),3)
 106  CALL PLOT(X(I1),0.5,2)
      I2=I1+1
      CALL PLOT(X(I2),0.5,3)
      CALL PLOT(X(I2),PROD(I2),2)
      CALL PLOT(X(III),CUMRES(III),3)
      KIII=III+1
      DO 8001 I=1,III
      KIIII=KIII-I
```

```
001 CALL PLOT(X(KIIII),CUMRES(KIIII),2)
    CALL PLOT(X(III),CUMPRO(III),3)
    DO 8002 I=1,III
    KIIII=KIII-I
002 CALL PLOT(X(KIIII),CUMPRO(KIIII),2)
    I2=III-1
    DO 8003 I=1,I2,2
    I1=I+1
    CALL PLOT(X(I),PROD(I),3)
    CALL PLOT(X(I),0.5,2)
    CALL PLOT(X(I1),0.5,3)
003 CALL PLOT(X(I1),PROD(I1),2)
    I2=I1+1
    CALL PLOT(X(I2),PROD(I2),3)
    CALL PLOT(X(I2),0.5,2)
000 CONTINUE
    CALL LASPLO
    STOP
    END
```

Chapter 3

The Future of Oil

— as seen by the Oil Industry in the late 1960s and early 1970s

The future shape of the global oil industry depends, as described in Chapter 1 and as specified in Chapter 2, on the inter-relationships of rates of increase in oil use, the annual rate of additions to reserves and the maintenance of an adequate R/P ratio, and the size of the world's ultimate oil resource base. As we shall show in Chapter 4, the contemporary conventional Western oil industry view on the future of oil is now acutely pessimistic. The industry appears resigned to the prospect of an early end to the world's continued economic development based on an increasing use of oil. It believes that the inevitable peaking of world oil production is in sight and alternative energy options have to be opened up as a matter of urgency. In Chapter 4 we try to show exactly what this view of the future of oil necessarily implies in terms of the values which have to be attached to the inter-related variables in demand, discoveries and resources. In this chapter we review the earlier opinions of the oil industry on the future of oil. More specifically we are concerned with views which the industry presented to world opinion and to energy policy makers in the West in a highly confident manner up to less than a decade ago. The contrast with its present gloomy prognostications is astonishing.

The literature of the oil industry in the late 1960s and the early 1970s provided a wealth of commentary on the exciting prospects which were then generally thought to present themselves to the industry, for the remainder of the century and beyond. Oil industry studies of that period, used in this chapter are essentially random choices from the wide range of literature available on future oil prospects. They are used to indicate the kind of future of oil that was then so confidently envisaged by the companies as they went about their daily business of discovering new oil reserves, of attempting to maximize their production of crude oil and of aggressively selling as much as possible of its derivatives, both of oil products for energy use and of chemical feedstocks. The purpose of the presentation in this chapter of the oil companies' view at that time is not only to enable us to pose the question as to why the industry has since so sharply changed its attitudes and conclusions about the future of oil, but also to enable us to analyse the implications of these views on the then expected growth rates in oil use for the question of the ultimate availability of oil. We thereby provide a basis of comparison with their current very different views – and those of others.

Up to the early 1970s oil consumption in the non-communist world was forecast by the companies as almost certain to rise to a level of 71 million barrels per day (b/d) by 1980 compared with under 37 million b/d in 1970. (= 7.5 per cent per annum increase in use 1970-80.) The prospects at that time of a total use of oil in the decade of the 1970s which exceeded the amount used in all the previous history of the industry (230 compared with 225 x 10^9 barrels) was viewed with complete equanimity – as the following quotation from an editorial in the *Oil and Gas Journal* indicates:

> World-wide oil is still in plentiful supply and recoverable reserves plus prospects for more indicate no shortage for years to come . . . we are in no apparent danger of running out of oil . . . Oil is taking up the slack that gas, coal and the atom have, for different reasons, created by falling short on their promise.[1]

Somewhat longer term estimates of the development of the 20-year outlook for the non-communist world's output of oil were made in 1970 by the Senior Vice-President for the Mobil Oil Corporation's Exploration and Production Division. Mobil Oil's forecast was for a 1990 level of oil output of 98 million b/d. This represents a 5.25 per cent per annum increase in the use of oil in the non-communist world over the period 1970-90 and an estimated use of 27 million b/d more in 1990 than the Oil and Gas Journal's estimate for 1980. It was, moreover, an estimated development which in the decade of the 1980s would have involved adding about 450 x 10^9 barrels to reserves. The Mobil Oil spokesman went on:

> In this case the undiscovered petroleum potential probably is adequate to meet the projected market demand for the next two

1. 'Petroleum and the Energy Crunch', an editorial in the *Oil and Gas Journal* for November 15, 1971.

> decades . . . our exploratory capability . . . is such as to enable us to face these demanding decades and their challenges with confidence and optimism.[1]

At about the same time another industry spokesman – this time the Head of Production in Royal Dutch/Shell's Exploration and Production Division – forecast the outlook for oil for the 30 years up to the year 2000. In the context of this long-term estimate of the future of oil Shell argued that there was 'no real reason' to doubt the continuation of exponential growth in energy consumption and 'no doubt that there is plenty of energy available to the world so that one should not yet be thinking in terms of an energy crisis'. The Shell spokesman did, however, go on to comment that the production of other energy sources was constrained. For example, he thought that 'nuclear electricity development would be inhibited by industrial capacity, costs and know-how and that the production of solid fuels would be constrained by competition from hydro-carbons'. Thus, he argued, 'the gap must be filled with hydro-carbons'. In the light of all these considerations the Shell view of the 30-year energy/oil future was that 'the non-communist world's total energy demand by the year 2000 would be 300 million barrels per day of oil equivalent'. Of this volume of energy use the assertion by Shell was straightforward: 'oil will then supply 170 million barrels per day'.[2]

The forecast was made as late as October 1972 – exactly one year before the beginning of the oil crisis – when Shell, along with Mobil Oil and the industry's *Oil and Gas Journal* were still supremely optimistic about the oil industry's ability to sustain the historic 10-yearly doubling rate in the consumption of oil. This question was a matter of the specific interest in a 1973 study on potential world crude oil supplies.[3] Its conclusion was as follows:

> Even if total world energy production continues to grow at a rate of up to 6 per cent per annum (compared with a rate of 5.2 per cent per year from 1945-72) . . . oil could further *expand* (our italics)

1. J.D.Moody (Senior Vice-President, Exploration and Production, Mobil Oil Corporation) in 'Petroleum Demands of Future Decades', *American Association of Petroleum Geologists Bulletin*, Vol.50, No.12, December 1970, pp.2237-45.
2. A.Hols (Head, Production Division of Exploration and Production Co-ordination, Royal Dutch/Shell) in 'Oil in the World Energy Context', *Economist Intelligence Unit International Oil Symposium*, London, October 1972. His estimate implies an average growth rate in oil use of 5.4 per cent per annum over the 30-year period.
3. J.D.Parent and H.R.Linden, *A Study of Potential World Crude Oil Supplies*, Institute of Gas Technology, Chicago, 1973.

its current (1972) 50 per cent share of the world energy market until about 1990 and then maintain it until after the year 2000.

Over the preceeding few years, the *Petroleum Press Service*,[1] which at that time presented the distilled views of the major oil companies, had similarly hammered away at the same theme. In 1965, for example, it reported on a paper given by D.S.Ion and W.Jamieson of British Petroleum at the annual conference of the British Institute of Petroleum. It noted the authors' conclusions that there was no doubt that future oil demands would be met in spite of the magnitude of the amounts required to 1980 'and well beyond'. The *PPS* went on to comment that 'the role of the oil industry is to develop the non-communist world's reserves efficiently'. It concluded its analysis by arguing, 'provided the potential demand is there to be satisfied, oil still undiscovered will be proved'.[2]

The following year the *PPS* discussed the papers given by various oil companies at the Tokyo meeting of the World Power Conference. In particular, it drew attention to another paper given by delegates from the British Petroleum Company (K.A.D.Inglis and W.Jamieson). This was entitled, 'Factors in the Adaptability of Petroleum to meet the Changing Structure of Future Energy Needs' and in it, according to the *PPS*:

> the authors gave cogent reasons why petroleum is particularly adaptable to modern needs and may thus be expected to play a major part in supplying the world's rising energy requirements *whatever structural changes may occur* (our italics) . . . Petroleum fuels are expected to remain competitive with all other sources of energy. Given reasonable freedom of the operation of the laws of supply and demand, *it can be asserted that there will always be enough petroleum for the world's needs*[3] (our italics).

Both editorial comments and reports on papers making much the same points continued to appear in the *PPS* in the late 1960s and up until 1973. For example, in October 1972 in an article which dealt critically with 'warnings of an impending energy crisis', it re-stated its position on the oil sector as follows:

> Oil demand has been doubling every ten years and world oil reserves are much smaller than reserves of coal. We shall not, however, run out of oil in the twentieth century . . . and a world shortage of fossil fuels is not in prospect for this period . . . . An evident need . . . is to press on with the search for fresh oil reserves to keep pace with the growth

1. Now renamed *Petroleum Economist* and published monthly in London.
2. *Petroleum Press Service*, Vol.XXXII, No.7, July 1965, p.258.
3. *ibid.*

in requirements.[1]

Indeed, it is seldom that one finds other than oil industry optimism in the literature of that period in respect of the oil industry's ability to continue to find, to develop and to produce the oil resources which would be needed to serve the world's rapidly growing needs. This conclusion was held to be valid even with the prospect of a doubling rate in oil use over every succeeding decade to the end of the century. The authors of the papers and reports quoted describe and explain, at length, how the industry would, if given the freedom to do so, keep up with the task imposed on it by a seven per cent, or higher, exponential growth curve in demand. Mr Hols of Royal Dutch/Shell, for example, wrote:

> The oil industry is busily engaged in searching for new sources of supply. Oil is becoming more difficult and more expensive to find but, nevertheless, advancing technology is opening up new areas to drill both on land and off-shore .... From the oil demand figures, one can derive, for different points in time, the resources required to satisfy demand and the inventories needed in order to guarantee future requirements .... If we are to provide the reserves needed... additions to ultimate recoveries of 40-45 x 10^9 barrels of oil per year will be required in years reaching to the end of the century. This would compare with an average of about 30 x 10^9 barrels per year in the last 15 years. Part of these additions will, of course, come from improved recovery techniques .... At present only about 30 per cent of the oil in a reservoir can be brought to the surface economically. An improvement of only one per cent in all known commercial fields would add 1½ years production to proven reserves at current rates .... [However], at least 50 per cent of additions to reserves will have to come from new discoveries . . . and this means a very intensive and continuous development effort. All this . . . need not come from new areas as there are a number of current production areas where large potentials could still be realized. Moreover, in the 1980s reserves from tar sands and shales should be contributing [to the supply].[2]

Such oil industry optimism on the large scale opportunities for expanding the annual additions to reserves and the annual rate of production was, it should be remembered, expressed when the price of a barrel of oil was less than $2 (in 1974$ terms – see Figure 1.1) and Hols, in his paper, certainly indicated the need for higher prices for oil in order to secure the required supply. He wrote in the following terms:

1. *Petroleum Economist*, Vol XXXIX, No. 10, October 1972, p.332.
2. Hols, *op.cit.*, pp.5-8.

> We should not subscribe to the modern day prophets who predict the exhaustion of energy reserves – and, in particular hydro-carbons – *in the foreseeable future* (our italics). The world contains large amounts . . . although the days of ample availability at today's price are over . . . the world will depend for a long time to come on hydro-carbons for its basic energy needs.[1]

In this paper Hols did not actually specify what price increases would be necessary to ensure 'ample availability' from the large amounts of oil which, he wrote, the world still contains. This, however, was spelt out by another Royal Dutch/Shell spokesman, Mr G.Chandler, just a few months later when he gave the 'keynote speech' to the 1973 annual summer meeting of the British Institute of Petroleum.[2] In this he indicated a possibility of the oil price having to rise to more than $7 per barrel by 1980 if the industry was going to be able to supply the amount of oil required – on the assumption that the then 7½ per cent per annum exponential growth in demand was to continue.

The industry's expectation that a price rise from $3 to $7 per barrel would provide sufficient economic incentive for the enhanced recovery and the intensive exploration effort necessary to find enough new reserves in both old and new areas to sustain a 7½ per cent rate of increase in demand has now to be seen in the light of a mid-1980 oil price – in 1970 $ terms – of more than $20 per barrel. In other words, actual price developments since 1973 have more than kept pace with the development the companies foresaw as necessary in order to ensure the adequate growth of oil reserves and of production potential for meeting the needs of the 'foreseeable future'.

Elsewhere in the literature one finds equally, or even more highly optimistic views on the prospects for oil supply/reserves developments. Parent and Linden, for example, in the 1973 study indicated a minimum estimate for ultimately recoverable oil of 3000 x 10^9 barrels and went on to say:

> of course, future recovery of crude oil in place may be increased . . . by major improvements in secondary and tertiary recovery procedure and estimates of the remaining resource base could also be revised upwards with further exploration and improvements in technology . . . . 4000 x 10^9 barrels may be taken as a second higher

1. Hols, *op.cit.*, pp.10-11.
2. G.Chandler, 'Energy: The Changed and Changing Scene', in Institute of Petroleum, *Energy from Surplus to Scarcity?*, Applied Science Publications, Beaconsfield, 1974.

value of economically recoverable crude oil to reflect this possibility.[1]

Within the limits of these two sets of reserves' figures the two authors indicated that maximum rates of the annual discovery of oil would have to be 72-82 x 10^9 barrels and 93-105 x 10^9 barrels respectively – compared with a maximum achieved up to the time they were writing of about 74 x 10^9 barrels. The annual rate of oil production in 1972 was about 18 x 10^9 barrels and they saw this level increasing to 85 x 10^9 barrels (with a lower resource base estimate of 3000 x 10^9 barrels) and to over 100 x 10^9 barrels (with a 4000 x 10^9 resource base).

Some years earlier, in 1965, the *Petroleum Press Service*[2] quoted the proceedings of the annual meeting of the British Institute of Petroleum in a discussion on ultimate oil resources. These proceedings indicated ultimate non-communist world potential of conventional oil at over 3650 x 10^9 barrels with the oil shales and tar sands of North America accounting for another 2630 x 10^9 barrels giving a total of well over 6000 x 10^9 barrels of future oil availability. In this context, the *PPS* went on, 'the oil industry should have no difficulty in meeting future demands'.

Indeed, if there was to be a future for the oil industry from 1970 onwards, given an expected continued rapid growth in oil use, there had to be considerable optimism concerning the resource base and a great deal of confidence in the industry in its ability to find and to prove new reserves each year. This is shown in Figures 3.1 to 3.10 in which the post-1973 future of the industry has been simulated on the basis of the model described in Chapter 2. In these simulations historic data has been used for the period up to 1973. Thereafter the use of oil is assumed to increase at 7½ per cent per annum for as long as the size of different resource bases allow, or for as long as an annual addition to reserves, required for the simulation, remains within a specified maximum.

The results of these 10 cases, in terms of the size and date of peak production, the peak annual addition to reserves (where this is not a constraint), and the dates of the discovery of the last set of reserves plus the year when declining production first comes back to the level of production in 1973, are brought

1. J.D.Parent and H.R.Linden, *op.cit.*, pp.2-6.
2. *op.cit., Petroleum Press Service*, Vol. XXXII

together in Table 3.1. This shows that in those cases for which the resource bases are assumed to be 2000 and 3000 x 10^9 barrels, the industry's growth period would not have lasted out the century. In the two cases with no limits to the annual additions to reserves (Figures 3.1 and 3.3) all oil resources would have been discovered by 1993 and 1999, respectively, with peak annual production occurring in 1993 (at 66.4 x 10^9 barrels) and in 1999 (at 79.0 x 10^9 barrels), respectively. Under the constraints of resource bases of this size and with the production decline curve as specified by the model, the annual level of production would have been back to the level of 1973 by the years 2009 and 2018 respectively. With an assumption of limits on the annual additions to reserves (Figures 3.2 and 3.4) the situation is little changed. Peak production (much lower, of course, than in the no-limits cases) occurs in 1991 and 2000 and the rate of increase in production is constrained below the nominal 7½ per cent per annum as early as 1982 and 1983. Peak rates of production do not approach anywhere near the oil industry's early 1970s estimates for the future levels of production – not even for 1980, let alone for 1990 or 2000!

Such prospects for the outlook for oil from the perspective of the early 1970s are simply not in keeping with the optimism of the industry at that time. There is, indeed, an essential incompatibility between the optimistic way in which the industry saw its future and the constraints to be faced in reality if there were to be no more than 3000 x 10^9 barrels of ultimate oil resources, and if the use of oil was to continue to grow at 7½ per cent per annum.

Thus the optimistic view of the future which the industry so firmly held at that time clearly depended on the world's ultimate oil reserves being more than 3000 x 10^9 barrels. Even with 5000 x 10^9 barrels of ultimate resources (as can be seen from Figures 3.5 and 3.6) and from the summary of the results of this iteration in Table 3.1, the impact of an exponential growth curve of 7½ per cent per annum would have led to peak annual production in the early years of the 21st century. Moreover, with a no-limits to reserves additions' assumption, and in order to achieve the pattern of use shown in Figure 3.5 for the rest of the period of the industry's growth, all 5000 x 10^9 barrels of ultimate reserves would have had to be discovered by 2006. In addition a formidable peak finding rate of 219 x 10^9 barrels would have been achieved in 1996. This rate is three times the highest rate achieved prior to 1973. Figure 3.6 shows

Table 3.1 *The future of oil from 1973 with a 7½ per cent per annum growth in use*

Figure No. of Iteration	Resource Base Size ($x10^9$ bbl)	No Limit or Limit on Reserves Additions	Peak Production		Peak Additions to Reserves		Date of Discovery of last Reserves	Date that Production falls back to 1973 level
			Vol. ($x10^9$ bbl)	Date	Vol. ($x10^9$ bbl)	Date		
3.1	2000	NL	66.4	1993	92.0	1984	1993	2009
3.2	2000	L	46.7	1991	40.0(L)	1982	2011	2016
3.3	3000	NL	99.0	1999	132.1	1989	1999	2018
3.4	3000	L	54.6	2000	50.0(L)	1983	2031	2034
3.5	5000	NL	165.6	2006	219.1	1996	2006	2029
3.6	5000	L	77.5	2002	68.0(L)	1990	2059	2060
3.7	8000	NL	267.7	2012	363.6	2003	2012	2038
3.8	8000	L	98.2	2010	92.0(L)	1990	post-2080	post-2080
3.9	11,000	NL	362.9	2017	485.5	2007	2017	2045
3.10	11,000	L	115.3	2016	110.0(L)	1990	post-2080	post-2080

(L) Limit on additions to reserves applies in these cases.

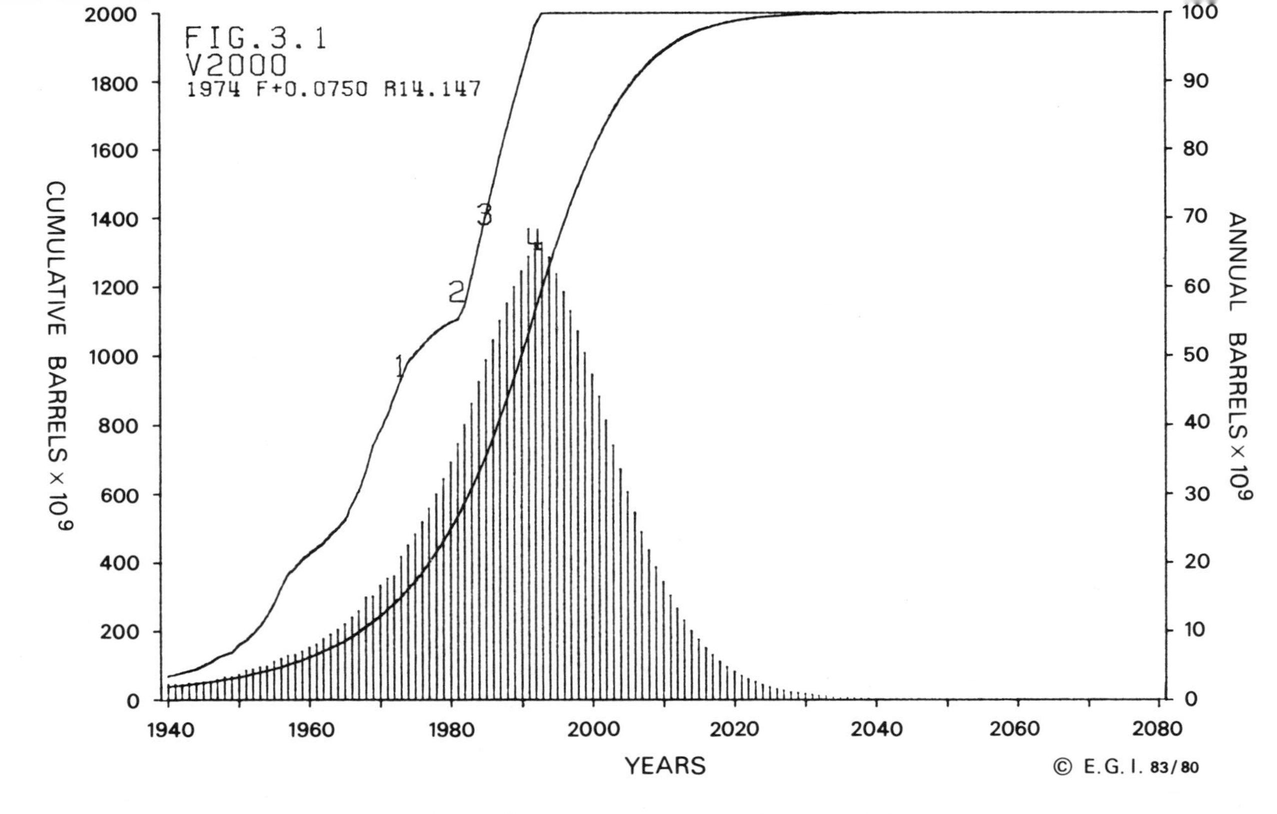

Figure 3.1 *The 2000 x 10^9 Barrels Resource Base, with a 7½% Growth Rate Case*

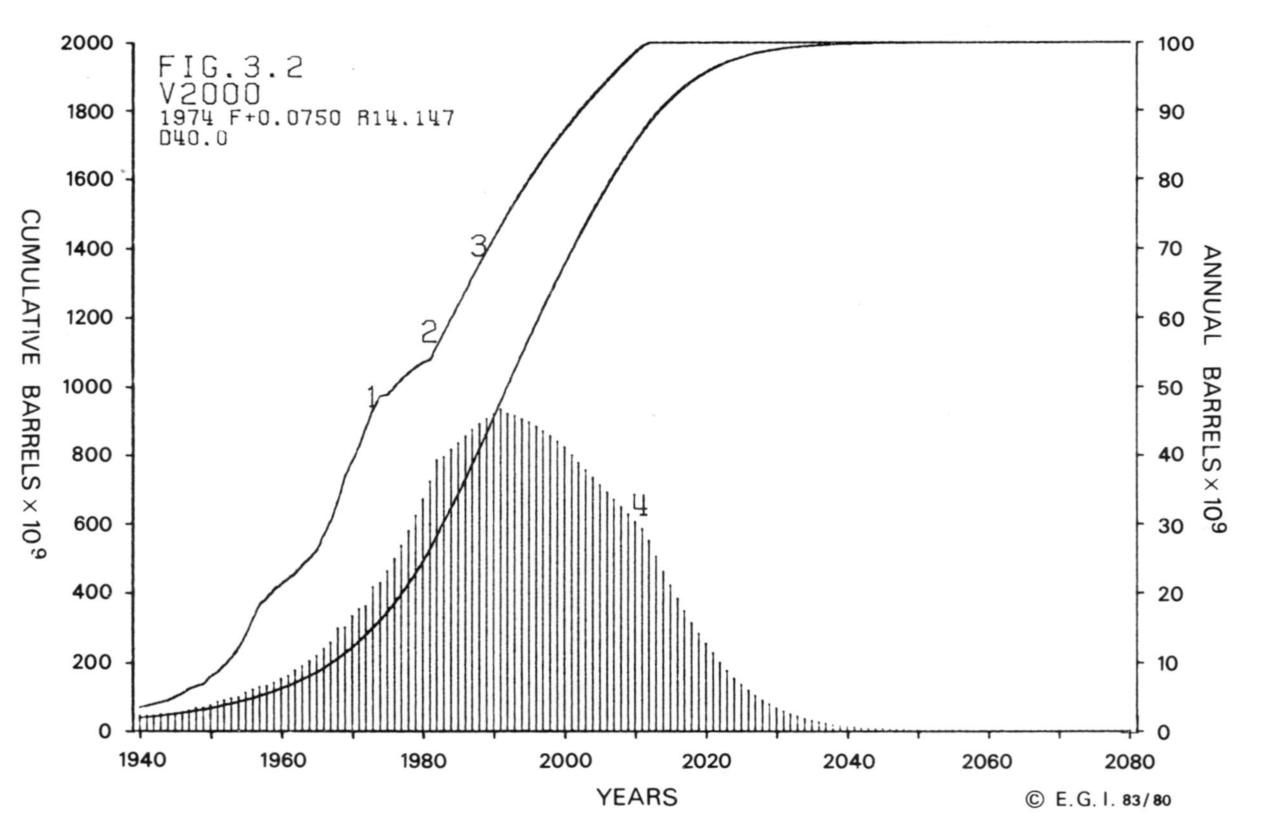

Figure 3.2 *The 2000 × 10^9 Barrels Resource Base with a 7½% Growth Rate Case*

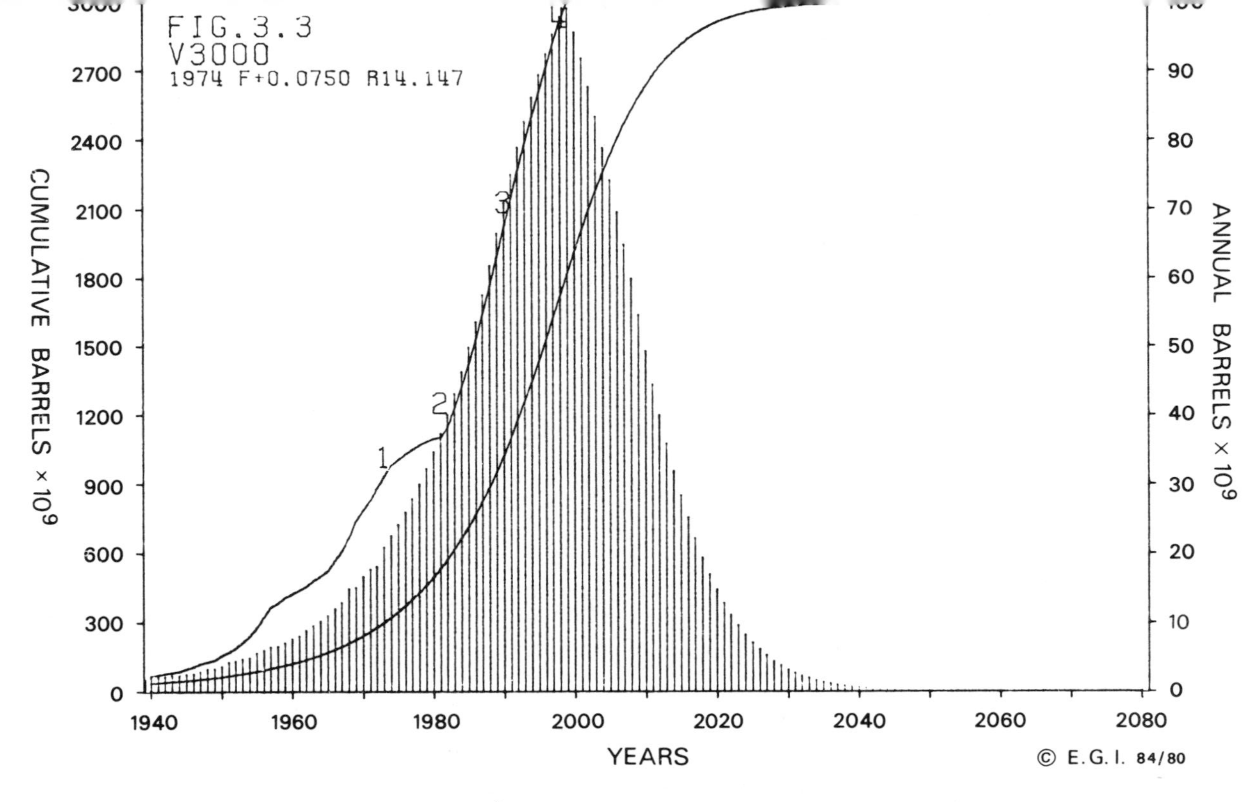

Figure 3.3 *The 3000 x 10^9 Barrels Resource Base, with a 7½% Growth Rate Case*

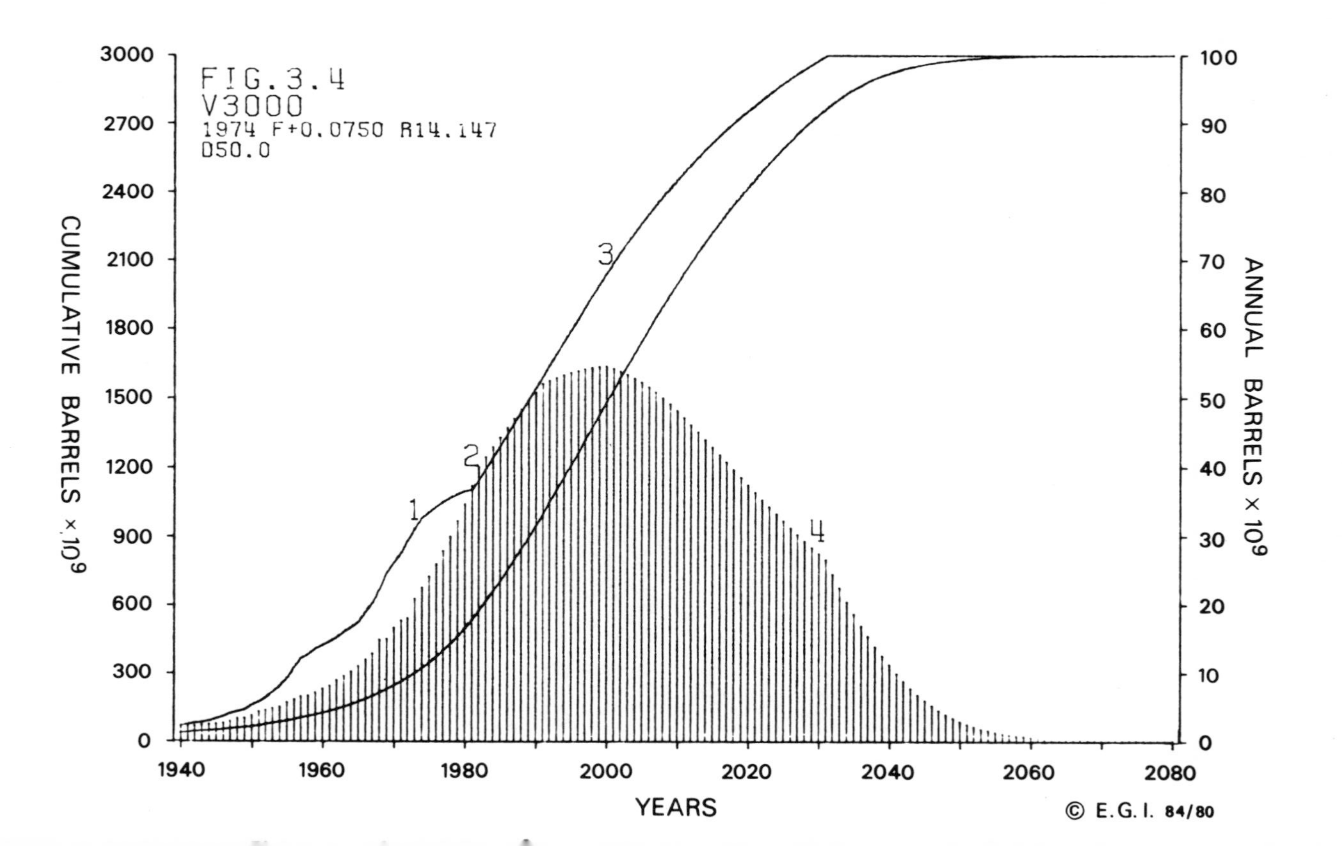
FIG. 3. 4
V3000
1974 F+0.0750 R14.147
D50.0
CUMULATIVE BARRELS × 10^9
3000
2700
2400
2100
1800
1500
1200
900
600
300
0
ANNUAL BARRELS × 10^9
100
90
80
70
60
50
40
30
20
10
0
1940
1960
1980
2000
2020
2040
2060
2080
YEARS
1
2
3
4
© E.G.I. 84/80

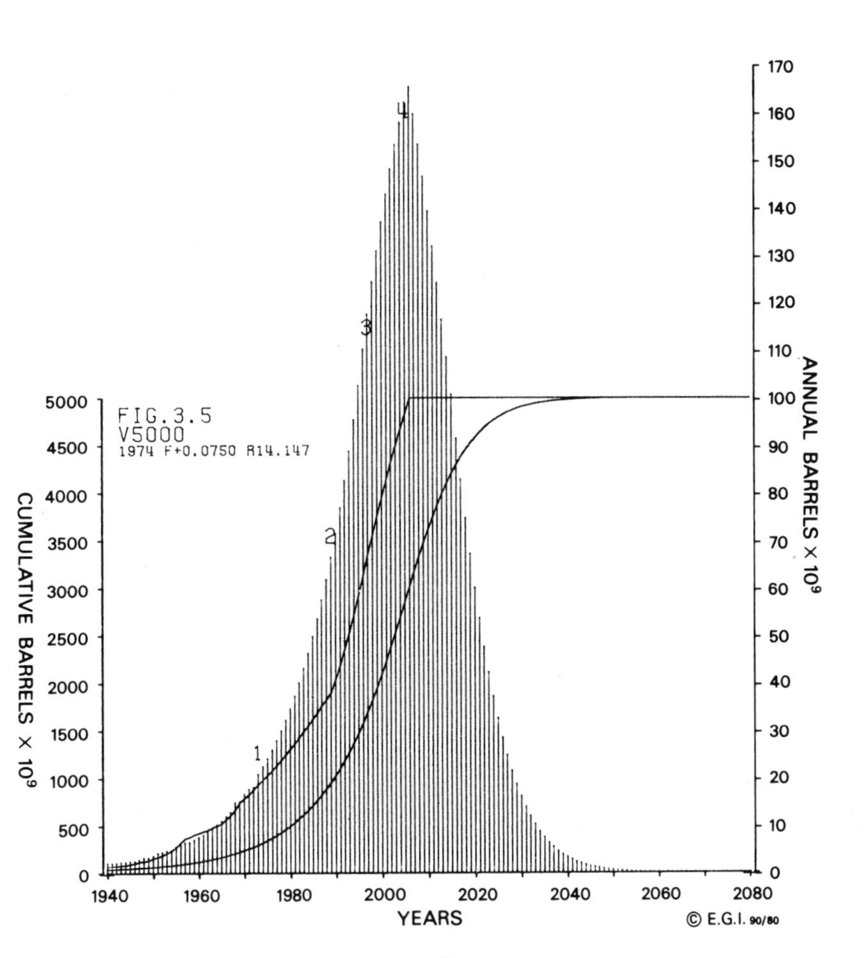

Figure 3.5 *The 5000 x 10^9 Barrels Resource Base, with a 7½% Growth Rate Case*

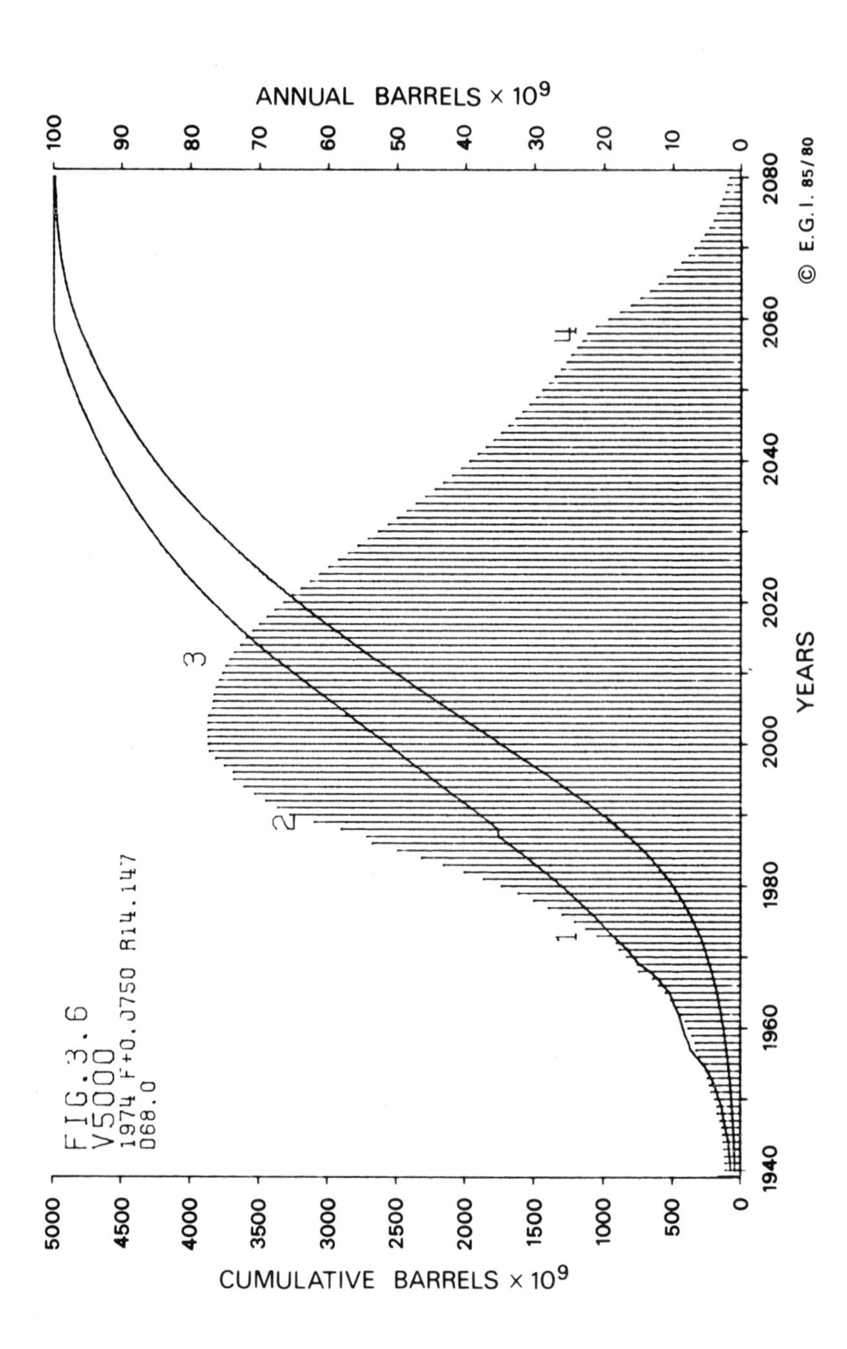
FIG. 3.6
V5000
1974 F+0.0750 R14.147
D68.0
ANNUAL BARRELS × 10^9
CUMULATIVE BARRELS × 10^9
YEARS
1
2
3
4
© E.G.I. 85/80

the impact of a limit of 68 x 10^9 barrels on the maximum annual additions to reserves. In this case the 7½ per cent exponential growth could only have continued to 1990. Thereafter, the growth rate in production becomes increasingly constrained by the limit of reserves additions so that the average annual growth rate from 1990 to the year 2000 is only 1.5 per cent. And, by the turn of the century, the level of production is only 77 x 10^9 barrels – less than the oil companies' early 1970s estimates for the level of output by 1980. In other words, the early 1970s expectation of oil industry spokesmen, that the industry could continue for the rest of the century in the same expansionist manner as it had done over the previous 25 years, depended on an ability to increase greatly the annual rate of discovery of reserves, even with a resource base as large as 5000 x 10^9 barrels. Without this, and even after allowing for a fall in the reserves/production ratio from the more than 30 years at which it stood in the early 1970s to only just over 14 years,[1] the industry could not have continued to expand at the rate then expected for more than another 16 years (from 1974-90). Yet, as we have seen, the oil industry confidently expected no problems before the turn of the century at the earliest.

With the higher resource bases of 8000 and 11,000 x 10^9 barrels and an unconstrained reserves' finding rate, the industry could continue to expand to 2012 and 2017, respectively: peak annual production levels would be 18 and 23 times the size of the industry in 1973 (see Figures 3.7 and 3.9). These projections required even higher annual rates of additions to reserves – of 364 and 485 x 10^9 barrels, respectively, and involved the addition of more reserves of oil in one year than all the oil used in the history of the industry up to 1973.[2] Figures 3.8 and 3.10 show the developments in reserves build-up and the production curves when there are limits on the annual rates of additions to reserves. These make little difference to the year of peak production but, in both cases, they make it impossible for the 7½ per cent exponential growth in the use of oil to continue beyond 1990. Even the limited maximum annual additions to reserves used in these iterations (92 and 110 x 10^9 barrels) are above the figures given for the future required rate of oil

1. A 14.2 years R/P ratio provides the oil which will be used over the following 10 years given the assumption of a 7.5 per cent exponential demand growth curve.
2. At the end of 1973 cumulative use of oil stood at about 300 x 10^9 barrels.

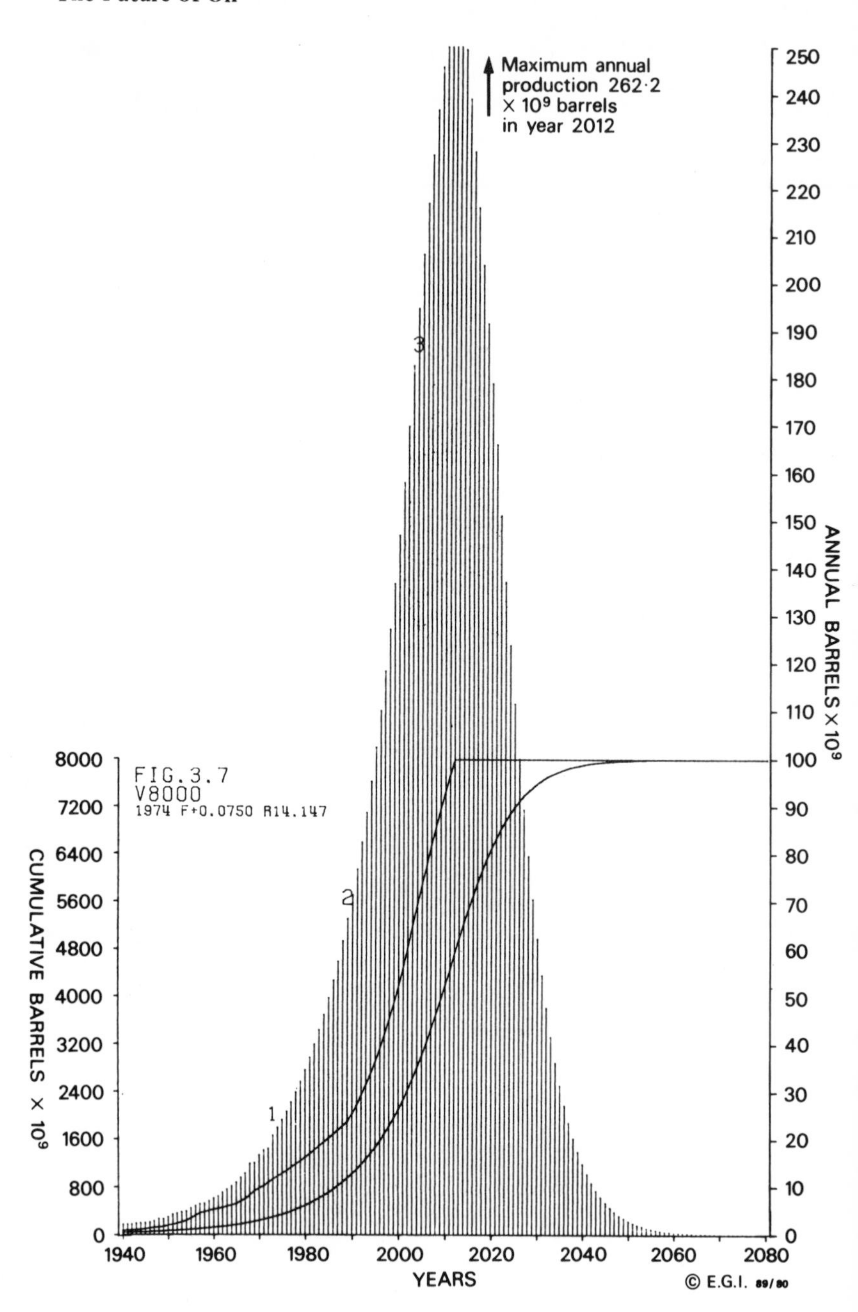

Figure 3.7 *The 8000 x 10^9 Barrels Resource Base, with a 7½% Growth Rate Case*

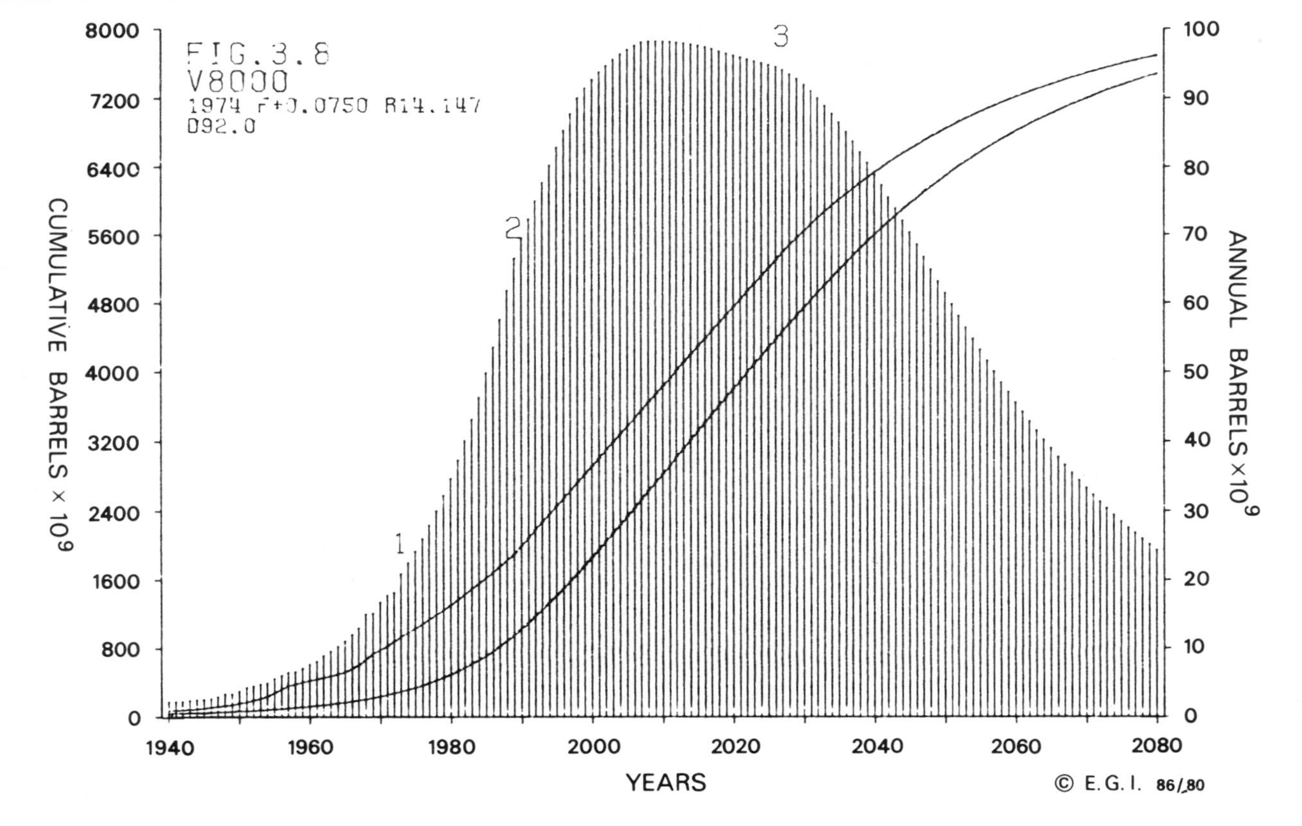

Figure 3.8 *The 8000 x 10^9 Barrels Resource Base, with a 7½% Growth Rate Case Additions to Reserves Limited to 92 x 10^9 Barrels*

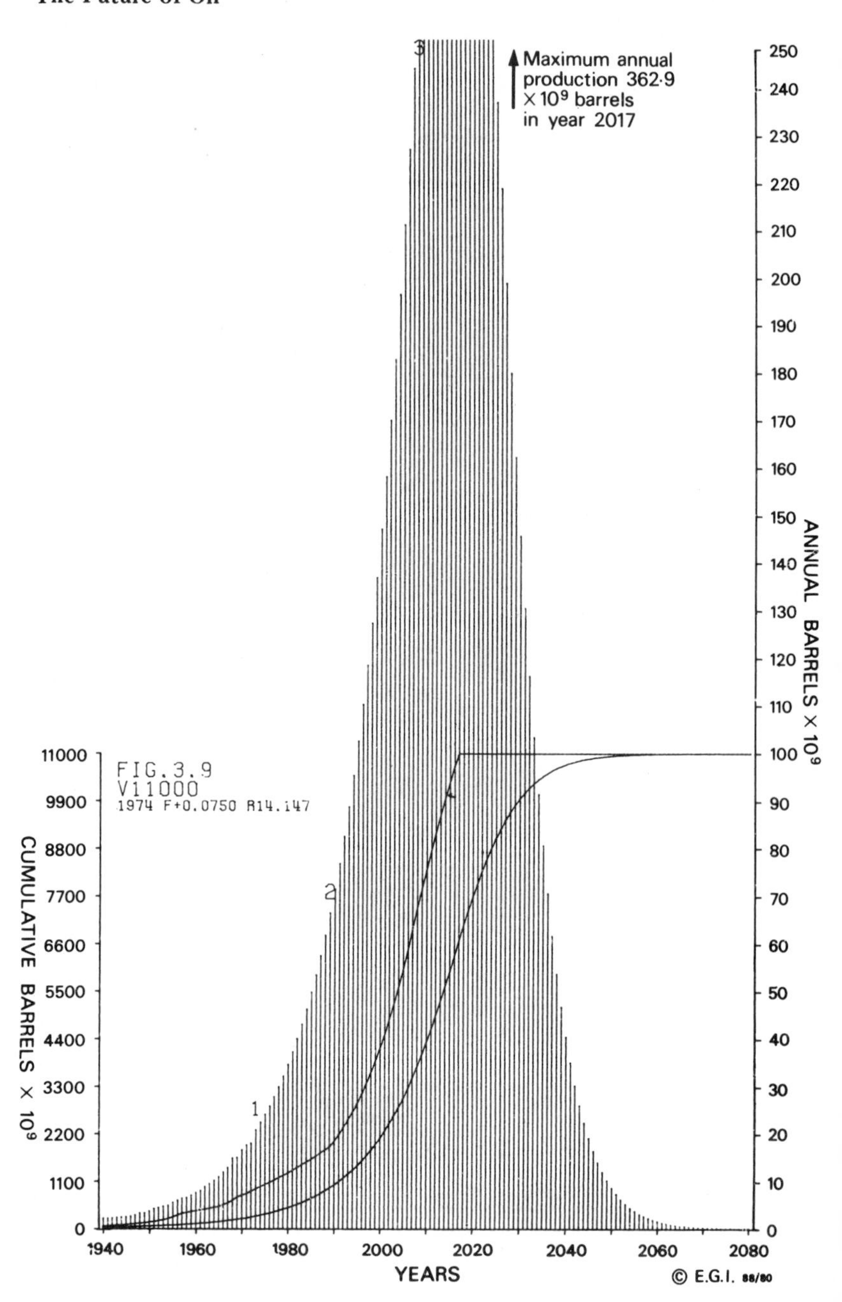

Figure 3.9 *The 11,000 x 10^9 Barrels Resource Base, with a 7½% Growth Rate Case*

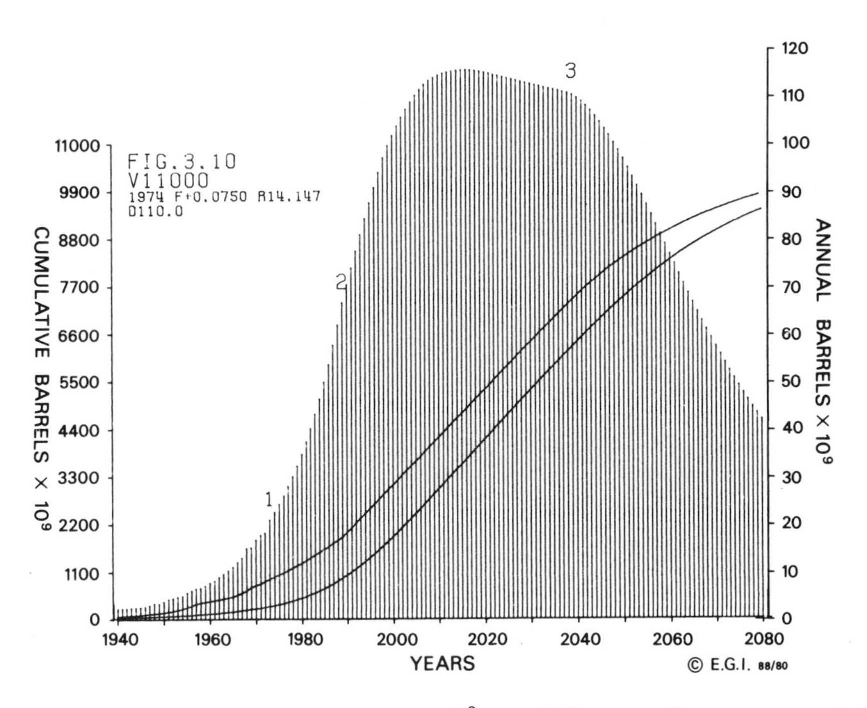

Figure 3.10 *The 11,000 x 10^9 Barrels Resource Base, with a 7½% Growth Rate Case Additions to Reserves Limited to 110 x 10^9 Barrels*

reserves additions by the oil industry's spokesmen in the early 1970s (see pp.96 to 97). Hols of Shell wrote of required additions to reserves of 40-45 x 10^9 barrels per year by the end of the century and Parent and Linden indicated a maximum of 105 x 10^9 barrels' figure for reserves additions.

Again the contrast between the industry's view of its future and the actual requirements of the expansion predicated, appear to show that one is justified in speculating on the bases for the oil industry's early 1970s view of the future. Its confidence in the long-term future of oil had to be based not only on an expectation of available resources far higher than the 2000 x 10^9 barrels, which the industry now says constitutes the most likely future, but also on an expectation of the ability of the industry to find vast new resources of oil each year (compared with what it had achieved to 1973) and that this would be accompanied by the large appreciation of already discovered reserves. This confidence must have been related to what were thought to be good prospects for technological advance in improving reservoir recovery rates. There was certainly no expectation in the industry at the time of a more than a 10-fold increase in the real price of oil within the period with which the industry's forecasts were concerned. The basis for the industry's optimism was not spelt out, but there is one element over which one can be confident: the continued historic rate of expansion of the industry necessitated an oil industry belief in an ultimate oil resource base of *at least* 5000 x 10^9 barrels. A base of at least this size did, indeed, give the prospect for a year 2000 production level of 170 million barrels per day – as forecast by Hols of Shell[1] – and for the industry to continue for some decades thereafter albeit in a declining phase of production.

Even this interpretation seems to be somewhat at variance with the industry's optimism over the long-term oil supply potential. It is possible that the oil industry did not really mean that the 7½ per cent exponential growth in the use of oil would continue for the rest of the century. Indeed, one of the oil industry's spokesmen, whose work was mentioned earlier in the chapter, Mr A.Hols of Shell, did specifically indicate in his paper that growth in oil demand could eventually be expected to moderate to about five per cent per annum.[2]

1. A. Hols, *op.cit.*, (see above p.94).
2. *ibid.* His overall growth rate for the period 1970-2000 was 5.4 per cent per annum (see p. 94). Mobil Oil's forecast for 1970-90 was an average 5.25 per cent per annum (see p. 94).

We have, therefore, modelled an alternative early 1970s oil industry view of the future of oil. This is one in which the oil use growth curve exponent remains at 7½ per cent from 1973 until 1985 and then changes to a lower rate of five per cent. The results of these simulations are shown in Figures 3.11 to 3.20 and the main output from the simulations are presented in Table 3.2.

The pattern of results is not dissimilar from that which emerges from the use of the 7½ per cent exponential growth curve in oil use. Again, neither the 2000 x 10^9 nor the 3000 x 10^9 oil resource base offers the prospect of growth for the industry through to the end of the century – irrespective of the unlimited/limited rate of annual additions to reserves variation (Figures 3.11 to 3.14). With 5000 x 10^9 barrels of resources (Figures 3.15 and 3.16) the industry can continue to grow until the year 2000 though, in both cases, constraints on the rate of increase in use below the five per cent written into these iterations begin to apply before or about this time *viz* in 1992 with a limit on reserves additions, and in 2001 with no limits. The period around the turn of the century would have marked the beginning of the end of the oil-based era so undermining the optimistic oil industry perspective on the longevity of its development and importance. With 8000 and 11,000 x 10^9 barrels of resources (Figures 3.17 to 3.20) the industry generally remains a growth industry into the first two decades of the 21st century. Even with these large resource bases, it should be noted, that with limits to annual reserves additions it would have become impossible to continue with the required five per cent per annum growth in the rate of oil use beyond 1992: thereafter a lower growth rate becomes necessary. It is, however, post the year 2080 (the final date of the period with which the model is concerned) before the last reserves in these iterations are discovered and it is also after that date that the size of the industry sinks back again to its 1973 level.

In brief, even with a fall in the growth rate over the last 15 years of the 20th century, the future of oil, as seen and as presented by the companies in the late 1960s/early 1970s, depended on the ultimate availability of resources at the upper end of the range of possibilities, and on the success of the companies in increasing the amount of oil added to the industry's reserves each year. By 1973 the international oil companies had 25 years of experience of rapid expansion under their belts; they had succeeded in achieving greater and greater

Table 3.2 *The future of oil from 1973 with a 7½ per cent per annum growth rate to 1985 and then a 5 per cent rate of increase in use*

Figure No. of Iteration	Resource Base Size (x10^9 bbl)	No Limit or Limit on Reserves Additions	Peak Production		Peak Additions to Reserves		Date of Discovery of last Reserves	Date that Production falls back to 1973 level
			Vol. (x10^9 bbl)	Date	Vol. (x10^9 bbl)	Date		
3.11	2000	NL	66.4	1993	92.0	1984	1984	2009
3.12	2000	L	46.0	1992	40.0(L)	1982	2012	2016
3.13	3000	NL	86.6	2003	106.7	1991	2004	2020
3.14	3000	L	54.1	2003	50.0(L)	1983	2032	2035
3.15	5000	NL	140.6	2012	173.7	2001	2013	2033
3.16	5000	L	75.6	2000	68.0(L)	1992	2061	2061
3.17	8000	NL	218.6	2022	269.5	2010	2023	2045
3.18	8000	L	94.7	2012	92.0(L)	1992	post-2080	post-2080
3.19	11,000	NL	293.0	2028	361.2	2016	2029	2054
3.20	11,000	L	113.2	2022	110.0(L)	1992	post-2080	post-2080

(L) Limit on additions to reserves applies in these cases.

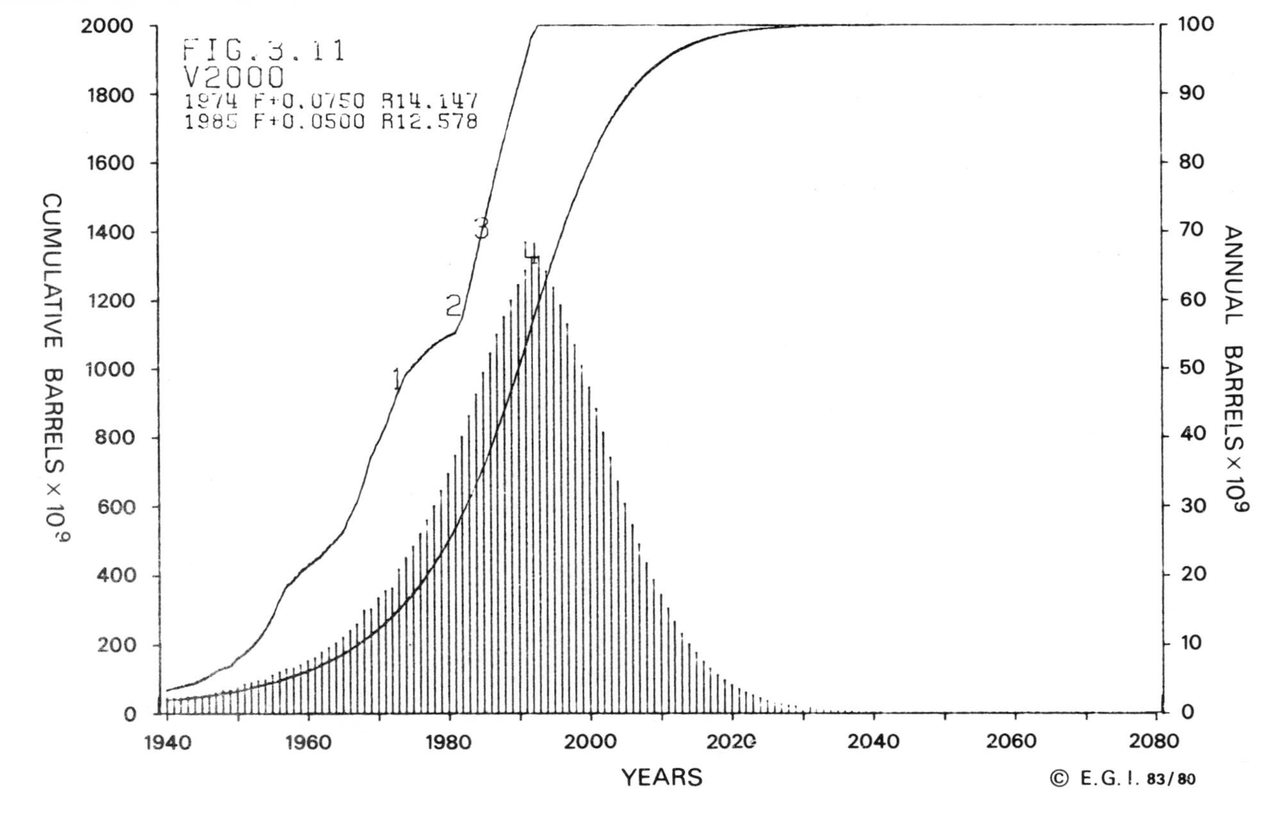

Figure 3.11 *The 2000 x 10^9 Barrels Resource Base, with a 7½%/5% Growth Rate Case*

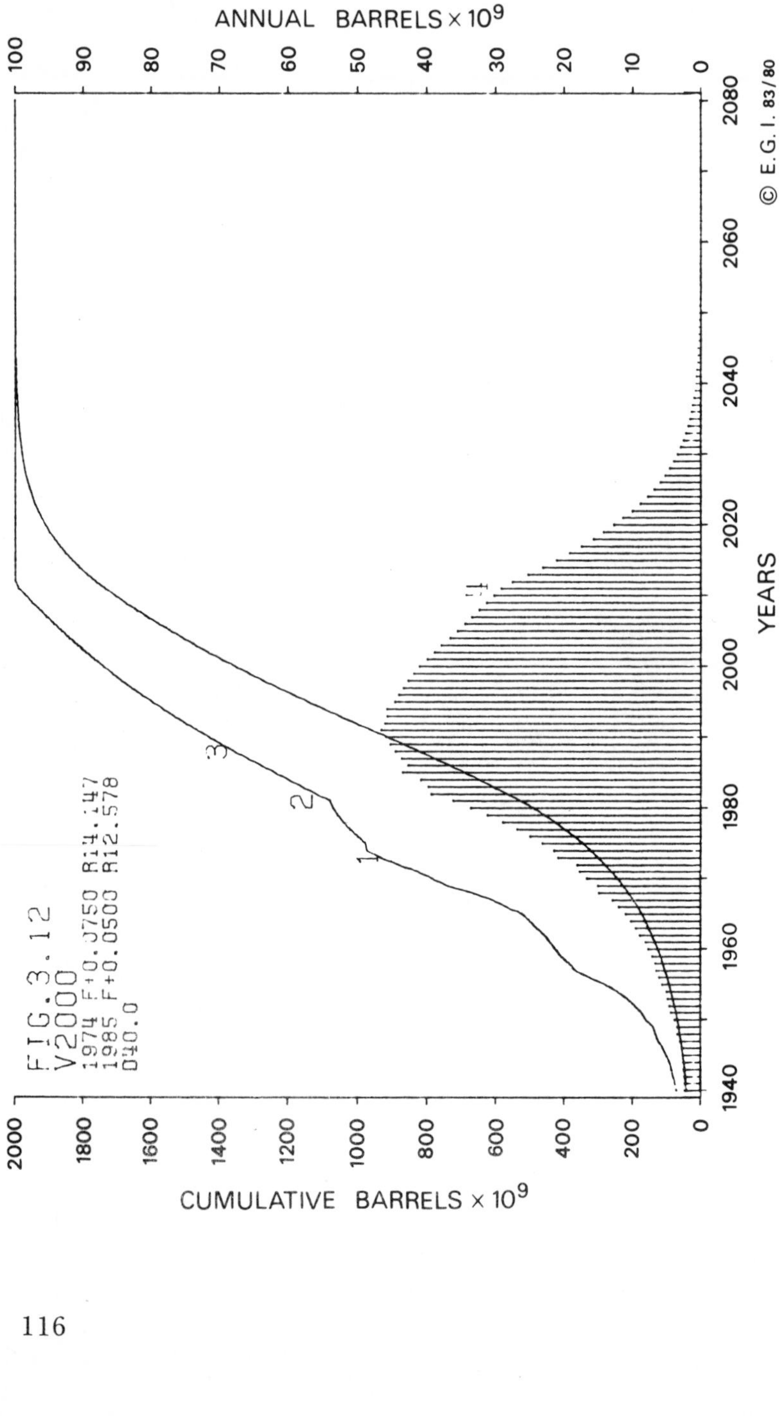

Figure 3.12 *The 2000 x 10^9 Barrels Resource Base, with a 7½%/5% Growth Rate Case*

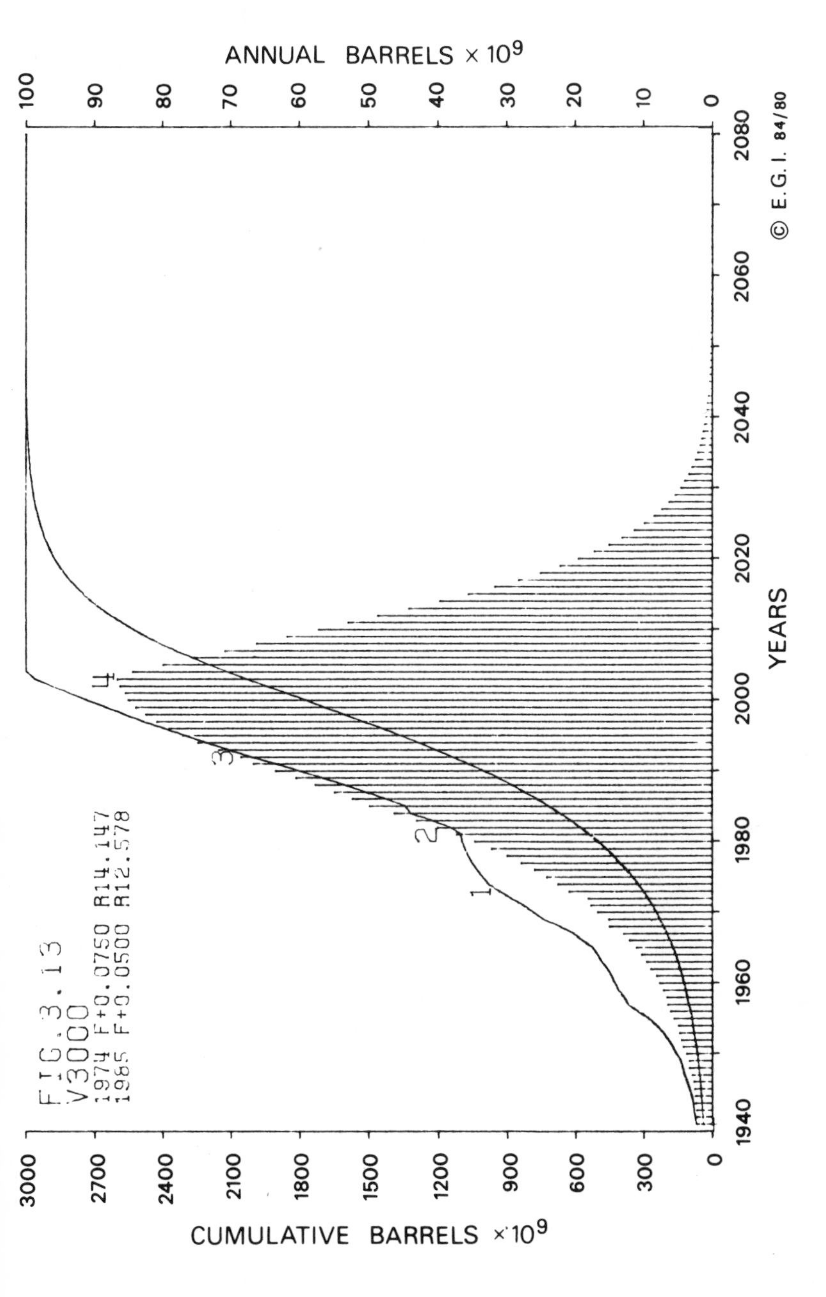

Figure 3.13 *The 3000 x 10^9 Barrels Resource Base, with a 7½%/5% Growth Rate Case*

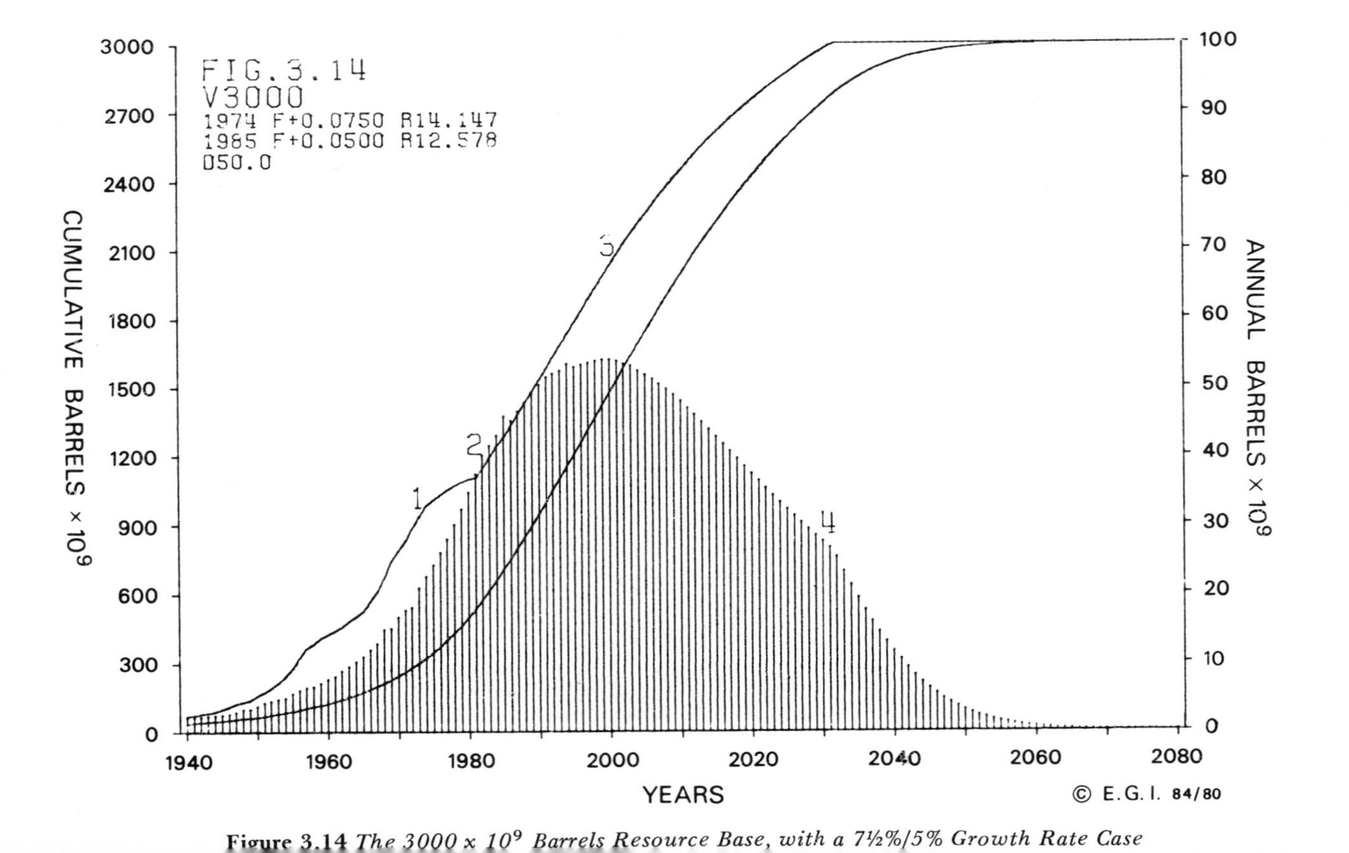

Figure 3.14 *The 3000 x 10^9 Barrels Resource Base, with a 7½%/5% Growth Rate Case*

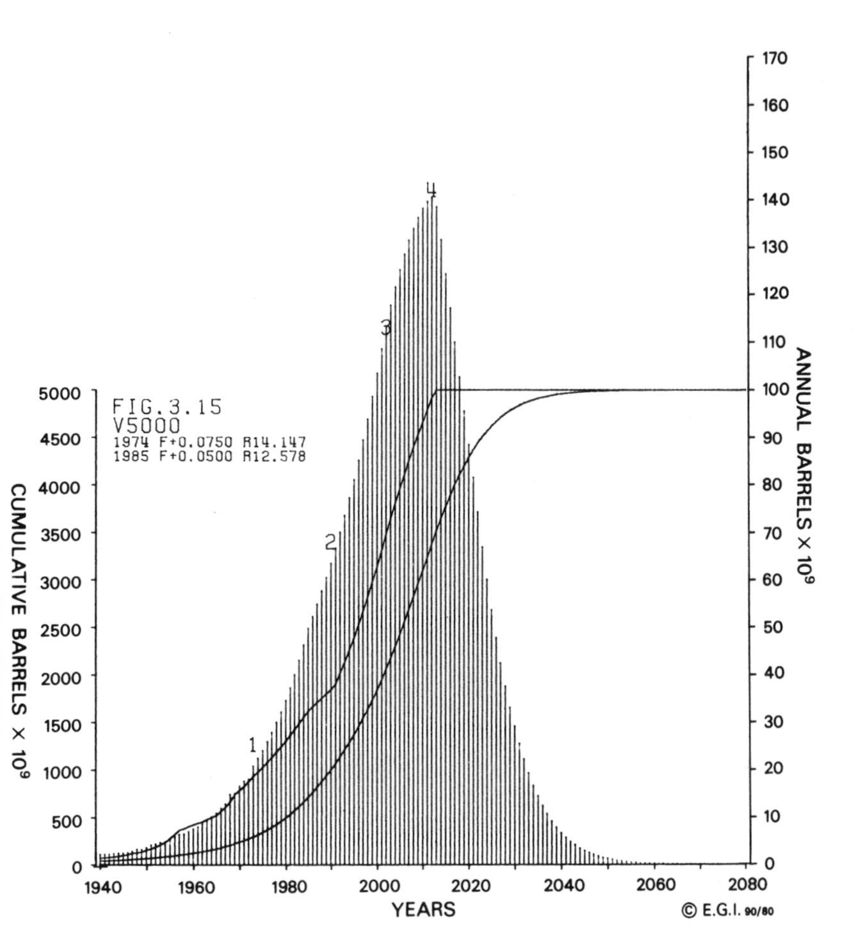

Figure 3.15 *The 5000 x 10^9 Barrels Resource Base, with a 7½%/5% Growth Rate Case*

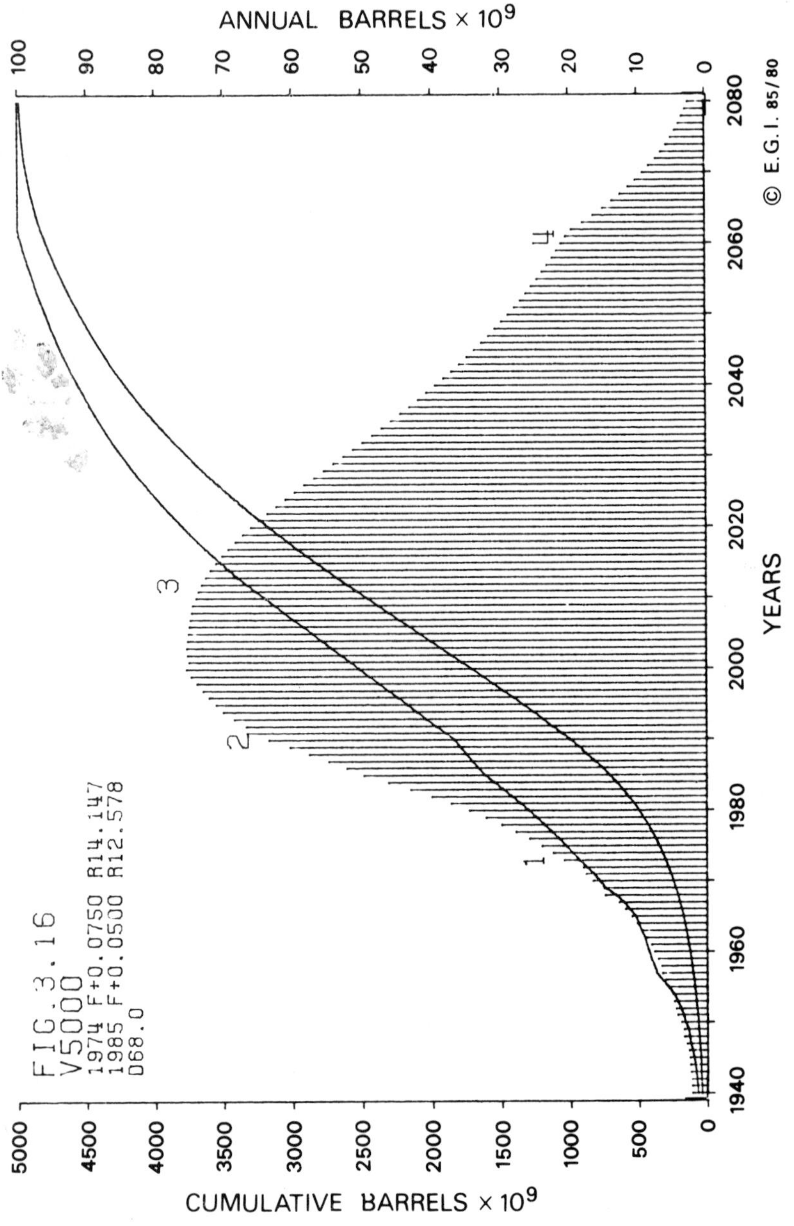

Figure 3.16 The 5000 × 10^9 Barrels Resource Base with a 7½%/5% Growth Rate Case

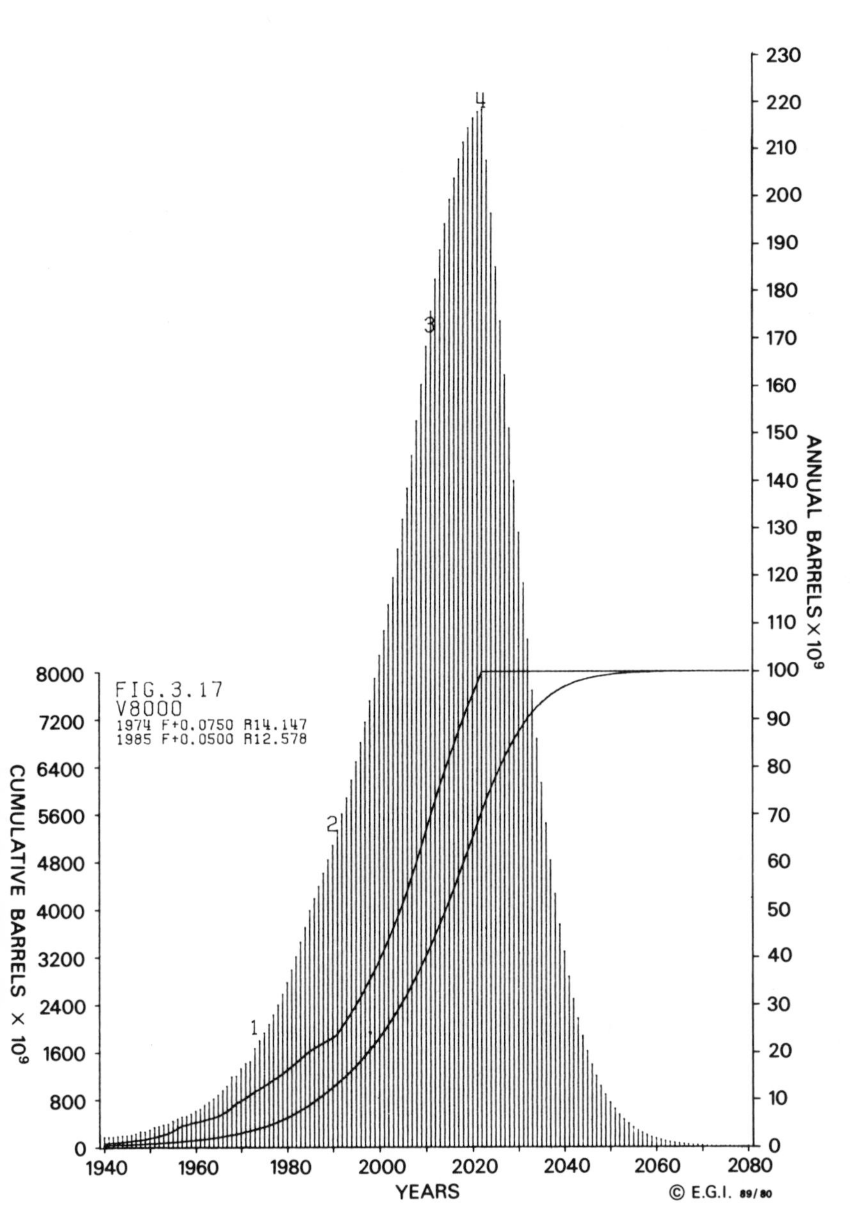

Figure 3.17 *The 8000 x 10^9 Barrels Resource Base, with a 7½%/5% Growth Rate Case*

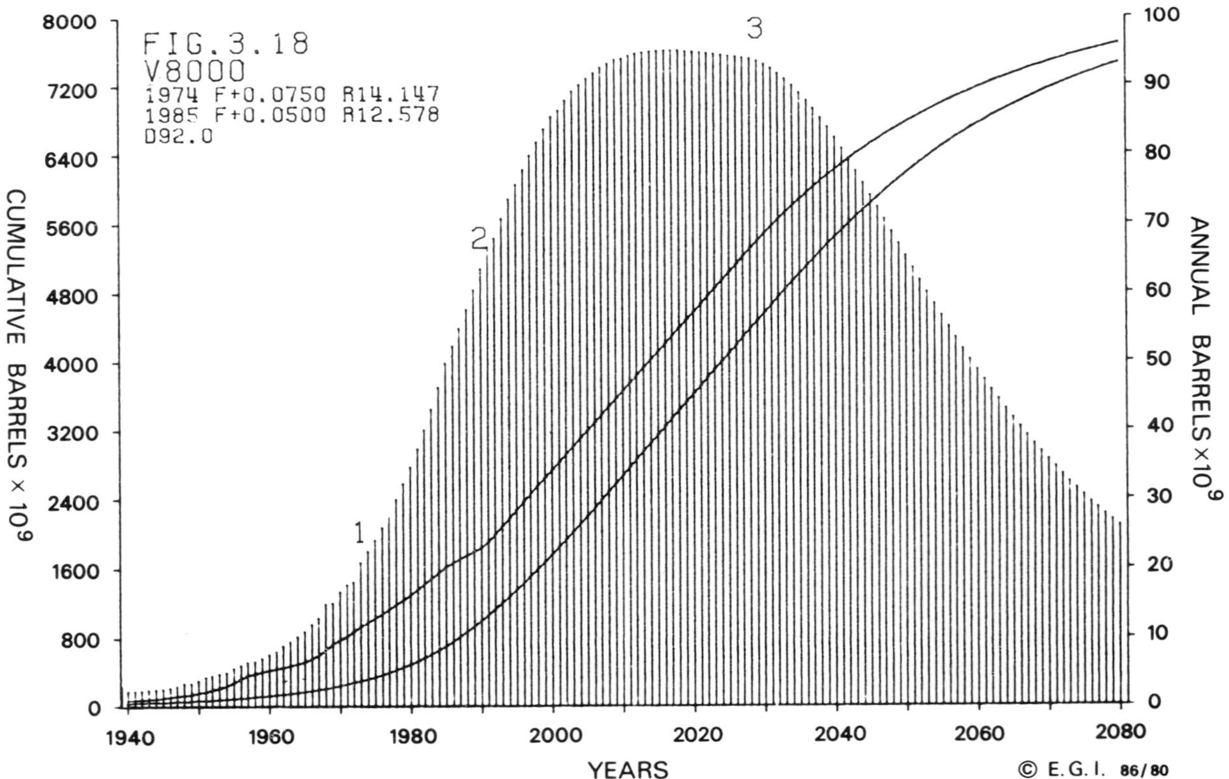

Figure 3.18 *The 8000 × 10^9 Barrels Resource Base, with a 7½%/5% Growth Rate Case*

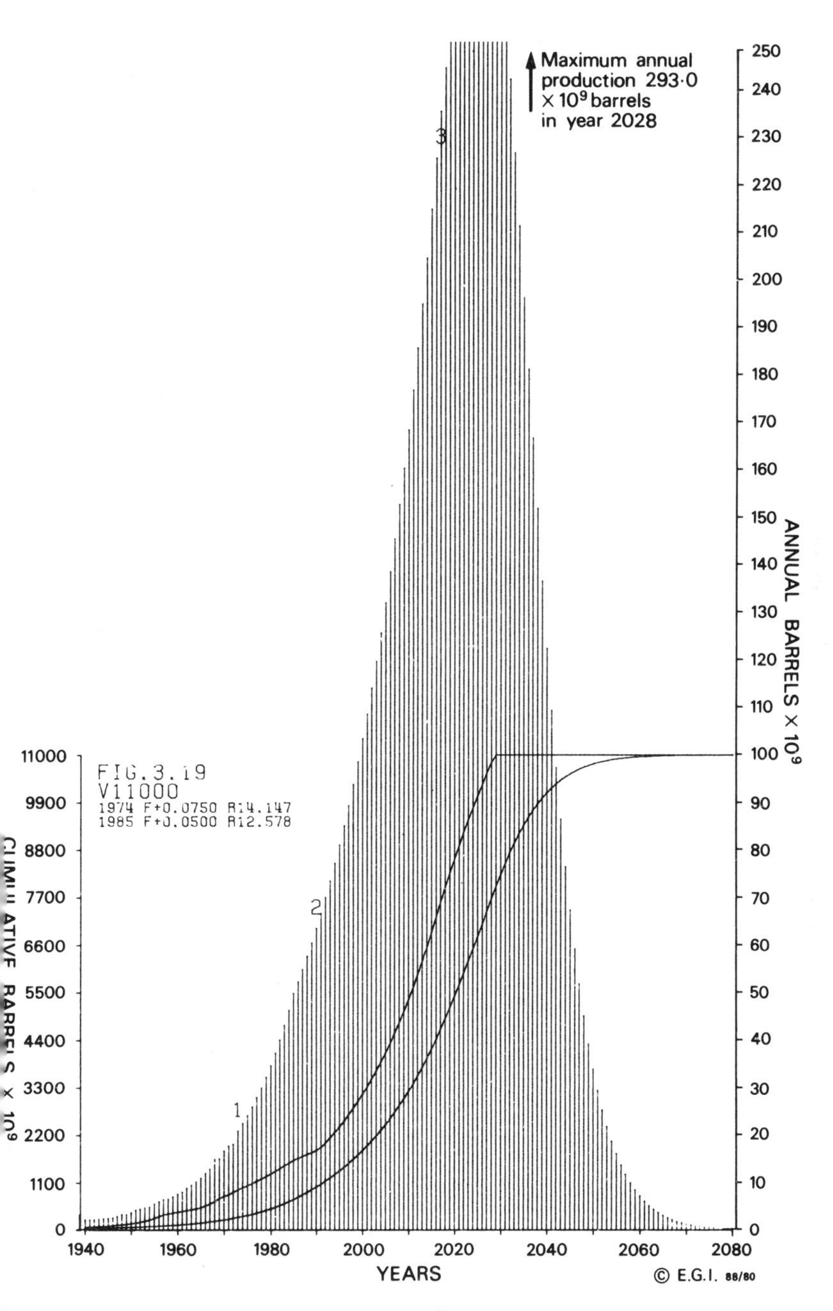

Figure 3.19 *The 11,000 x 10^9 Barrels Resource Base, with a 7½%/5% Growth Rate Case*

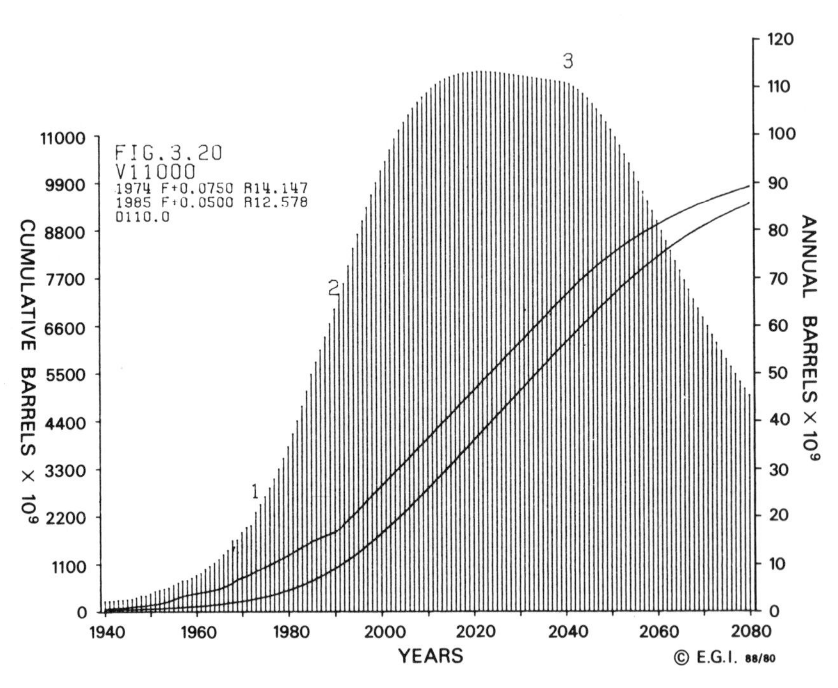

Figure 3.20 *The 11,000 x 10^9 Barrels Resource Base, with a 7½%/5% Growth Rate Case Additions to Reserves Limited to 110 x 10^9 Barrels*

economies of scale in most of their activities; they had developed their technology of oil search and of oilfield development to levels of sophistication which augured well for the future, both in respect of improving recovery rates in mature areas of production and of achieving success in new areas of oil potential – as these became necessary to develop along with growing demand. Moreover, by the early 1970s the lean years financially (as shown in Figure 1.2) appeared likely to be over. Thus with the expectation of higher rates of return on investment to cap their scientific and technological optimism, it is not unreasonable that the companies should view the future so full of hope. But lying behind all this, as the essential element from which their optimism emerged, was their unconsidered view of a world so devoid of additional oil resources that the growth rates, to which the industry had become accustomed, could not be sustained. As far as the companies in the late 1960s and the early 1970s were concerned, the expansion of the industry's resources to whatever levels were required was essentially a function of their continued investment in the search for oil and their confident expectation that they would not only find more new petroliferous regions and more new fields in existing regions, but that recovery rates would steadily improve through both technological developments and the impact of improved economics: in a world which, though politically complex, was nevertheless one in which, in the final analysis, the companies were acceptable as instruments whereby the oil required for continued economic development could be obtained.

Optimism, confidence and certainty in the future of oil were the hallmarks of the oil industry a decade ago; and to this must be added an element of unreality concerning the implications of their stated views in the light of the resources and reserves situation. The contrast between that outlook and the view which now prevails will become apparent in the following chapter in which the irrationality of the companies' present views will also be made clear.

Chapter 4

Presentation of the Oil Industry's New Conventional Wisdom

A. Introduction

The public pronouncements of the oil industry on the future of oil have recently undergone a traumatic transformation. In place of the former supreme optimism there is now the presentation of an essentially sombre prospect as far as the future availability of oil is concerned. This sombre, even somewhat frightening view of the future is captured in the contents and messages found in recent newspaper advertisements by Shell and British Petroleum, reproduced here in Figures 4.1 and 4.2. They state, in essence, that the world stands at the beginning of the end of the oil era in the process of economic development and that there is a need to take immediate action to ensure the availability of alternative energy sources. These, it is argued, are required to take over from oil in the relatively near-term future.

The contrast between these views and earlier ones held by the oil companies are startling. The impact of the apparent radical change of mind has been equally formidable. The earlier comfortable view of the future of oil – a view effectively sponsored by the oil companies themselves – has now been replaced by an all-round concern, or even fear, for the future. This is presented not simply in terms of an insufficiency in the future supply of oil, but also in terms of consequential much broader issues in international relations and policy-making such as Western concern for the region of the Gulf, where so large a percentage of the world's currently known oil reserves are concentrated.[1]

1. See Chapter 6 for a further consideration of such policy issues and the implications of the fear of scarcity of oil syndrome on policy-making by the Western world.

The New Energy Age is on its way

What are we doing about it?

The energy debate is really just beginning. Theories abound, misconceptions are many. The only thing that is certain, is that anybody who makes firm predictions about the future will be in some part wrong. For now, the energy experts views' (and they are all upheld by Shell's own work on the subject), can be summed up like this:

1. Demand for energy will continue to grow in spite of price pressure. Even if price, government policies, and other factors force low growth in demand, supply will fall short in the next twenty years.
2. The shift from a world economy dominated by oil must begin now. Alternative sources of energy must be developed with utmost vigour. Coal and nuclear power will be the most important sources of future energy. 'Geophysical' (wind, tidal, solar etc.) sources are still far away.
3. In the UK, North Sea Oil provides welcome medium-term security of supply but does not mean cheaper energy for Britain.

COMPARATIVE ENERGY COSTS

Middle East Oil
North Sea Oil
Indigenous Coal (US)
Imported Coal (N.W. Europe)
Indigenous Coal (N.W. Europe)
Nuclear input break even value
Low BTU Gas from indigenous Coal (US)
Natural Gas imports
Natural Gas from indigenous Coal
Liquids from Coal / Oilsands / Shale
Liquids from imported Coal
Biomass (crops grown for fuel)
Solar hot water

$0 $10 $20 $30 $40 $50 $ barrel of oil equivalent

The New Coal Age

Coal, the energy source of the first industrial revolution, will be important again during the second. The world has known reserves of coal sufficient for 235 years, and the UK for 300. However, before coal can take its place in the New Energy Age, new technology is needed. Compared with oil, coal has inherent disadvantages. It is hard to transport and handle. It contains a high proportion of contaminants and supplies less energy per unit of weight. So, as well as investing in coal mining rights, Shell is developing the new techniques which will make coal more economical to use. One project is the development of a fuel called *Colloil*, which is a mixture of oil and very fine coal that could be shipped like oil and used in many modified oil-fired installations.

Another Shell project, in test with the NCB, is the 'pelletizing' of coal dust into tiny, regular-shaped pieces. These could not only save a great deal of coal now wasted as dust but could be pipelined and carried in tankers like liquids.

Another Shell test plant in Holland is perfecting a new method of coal gasification which will bring coal gas down to a cost comparable with gas made from oil. This could help to save the millions of tons of oil now turned into gas, for other uses.

WORLD ENERGY RESOURCES: COAL AND THE REST

URANIUM (90)*
NATURAL GAS (48)
OIL (30)
COAL (235)
BRITISH COAL (300)

No. of years production at current levels.

* Estimated assuming current nuclear electricity generating capacity and technology.

New oil from old wells

It is not generally known that only around 40%–45% of the oil in a well is recoverable. In a new well, oil flows under natural pressure. But when the pressure drops or the oil that remains is too thick to flow, secondary and even 'tertiary' methods have to be used to coax the oil from the well. To increase production from old wells, they can be flooded with water or injected with air or steam.

In some wells, detergents can be used to loosen viscous oil, or chemicals used to emulsify it.

SECONDARY RECOVERY IN AN OIL WELL

(A)
(B)

(A) Hot water injected into oil-sand (B) Oil-sands heat up, oil flows easier.

Free power from nature?

Dreams of unlimited free power from the sun, wind and tides may come true in the next century but for the foreseeable future, they will contribute little to world energy. Nevertheless, Shell is engaged in research into some forms of solar energy conversion and storage, and experts do not rule out a breakthrough which could change the picture dramatically.

SAVE IT

Save it

All fuel economists urge vigorous energy conservation measures *now*, as a means of easing the transition from oil to other fuels. For these and ordinary 'housekeeping' reasons, Shell itself gives conservation measures a high priority. Over the last three years, for example, flare gas recovery techniques, improved process control, improved tank insulation and other measures taken at UK refineries are now saving 40,000 tons of refinery fuel a year.

'People working with energy'

Although Shell is the world's third largest free-enterprise industrial organisation, its real wealth is in people and their ideas. Over 6,500 of these people work in research – most of it for practical application, but some of it fundamental to chemical, physical and biological science. Only through the energy and originality of Shell *people* does Shell hope to find solutions to energy problems.

Biological and chemical research at Sittingbourne – one of Shell's many research centres.

You can be sure of Shell

Figure 4.1 *Shell Oil Newspaper Advertisement*

We regret that British Petroleum refused permission for this illustration, a BP advertisement, to be reproduced. No reasons were given.

The advertisement 'We're working to help solve the world's energy problem' which was published in a number of British newspapers and journals in 1979, showed the proverbial candle burning at both ends. In the accompanying text BP indicated its belief that unless the use of energy was cut down the world would be short of oil and other forms of energy by the mid-1980s. Such views are not only extremely pessimistic in the context of contemporary supply/demand conditions but are also completely at variance with earlier views held by the oil companies, including BP. (See Chapter 3)

Figure 4.2 *British Petroleum Newspaper Advertisement*

In the light of the oil companies' complete *volte face* over the future of oil and the manner in which their slightly earlier views have been so conveniently forgotten and their latest views so generally accepted – even to the extent of governments exclusively basing their energy policy decisions on them – it is appropriate to look much more closely at what it is the oil companies are now forecasting as the future of oil. In particular, it is necessary to assess their forecasts from the point of view of their implications for the question of the world's oil resource base. This essential component in any consideration of the future of oil, is, as we have shown in Chapter 1, the subject of a wide range of estimates as to its value. The companies, as discussed in Chapter 3, have hitherto attached high values to it, as an inevitable part of their optimism about the future of oil. Their revised approach to the issue, in their new conventional wisdom about oil, is thus of great importance in judging the validity of their now pessimistic conclusions. Three readily available and/or widely distributed oil industry (or industry supported/dependent) studies have been analysed and the results of the analyses are presented in the next three sections of this chapter.

B. British Petroleum's *Oil Crisis . . . Again?*

In September 1979 British Petroleum published a report of its Policy Review Unit (the company's 'Think Tank') under the title, *Oil Crisis . . . Again?*.[1] Though not formally an official company document it was accompanied by a statement by Mr Robert Belgrave, Policy Adviser to the Board of the British Petroleum Company Ltd. This statement was made at the press conference called to launch the attractively presented publication which, with a glossy and eye-catching cover, was clearly intended to have a wide circulation[2] and to provide as public a view as possible of BP's opinions. In this statement Belgrave said:

> We have for years been predicting that at some date in the future oil supply would cease to grow while the desire to consume it would continue to increase. That day may now have come. I do not believe that it would be prudent for the non-communist world (the NCW) to

1. British Petroleum, *Oil Crisis . . . Again?*, London, 1979.
2. World-wide circulation is, indeed, being attempted by the publication by local BP affiliated companies of translated versions of the report. See, for example, *Alweer een Oliecrisis?*, published by British Petroleum Maatschappij Nederland BV, Amsterdam, May 1980.

> count on there ever being more oil available to it than in 1978 . . . . Supplies in the NCW at or near the present level could be sustained well beyond the end of the century provided exploration and development are allowed at the rates of which the industry is capable.[1]

As things turned out Belgrave was soon to be proved wrong in terms of his opinion that the non-communist world 'could not count on there ever being more oil available to it than in 1978'. In 1979 the non-communist world had over four per cent more oil 'made available' to it than in 1978 – in spite of the severe cut-back in production in Iran from 1.99×10^9 barrels in 1978 to only 1.12×10^9 barrels in 1979.[2] In his remarks, however, Belgrave appears to have been attempting to highlight, in a simple effective way, the imminent onset of the beginning of the end of the oil era. He would, no doubt, maintain that production level changes of say ± four per cent from that of 1978 make no difference to the crux of his argument. Indeed, the results of the study which he was introducing indicate a somewhat different view of the short-term future of oil. It has non-communist world production increasing at a rate of approximately 2.76 per cent per annum from 1980 to 1985 when the peak rate of non-communist world oil production is shown, by the study, as being most likely to occur. This British Petroleum view of the future of oil is illustrated in Figure 4.3 which is based on illustrations presented in the report itself.[3] Thereafter, beyond 1985, according to BP, non-communist world production would decrease by about 1.94 per cent per annum until the end of the century. This is the average annual rate of decline for the 15-year period and it has, again, been calculated from the illustrative material in the Report. However, as Figure 4.3 shows, the trend is not a straight-line for the whole period so the use of the annual average rate does introduce a small error.[4]

1. From the introductory remarks made by Mr Robert Belgrave at a press briefing held at Britannic House, London to mark the publication of the paper, *Oil Crisis . . . Again?*, September 27, 1979.
2. Non-communist world production in 1979 was 18.43×10^9 barrels compared with 17.75×10^9 barrels in 1978. In addition the communist world (of USSR and China) also increased their exports to the non-communist world to a record level in 1979.
3. British Petroleum, *op.cit.* The illustrations involved are Figure 20 and its overlay. These cannot be reproduced here because of technical considerations, *viz* the use of colour in the original and the complexity introduced by the semi-transparent overlay.
4. Unfortunately, the text of the paper does not give the data used for the presentation nor does it explain its derivation.

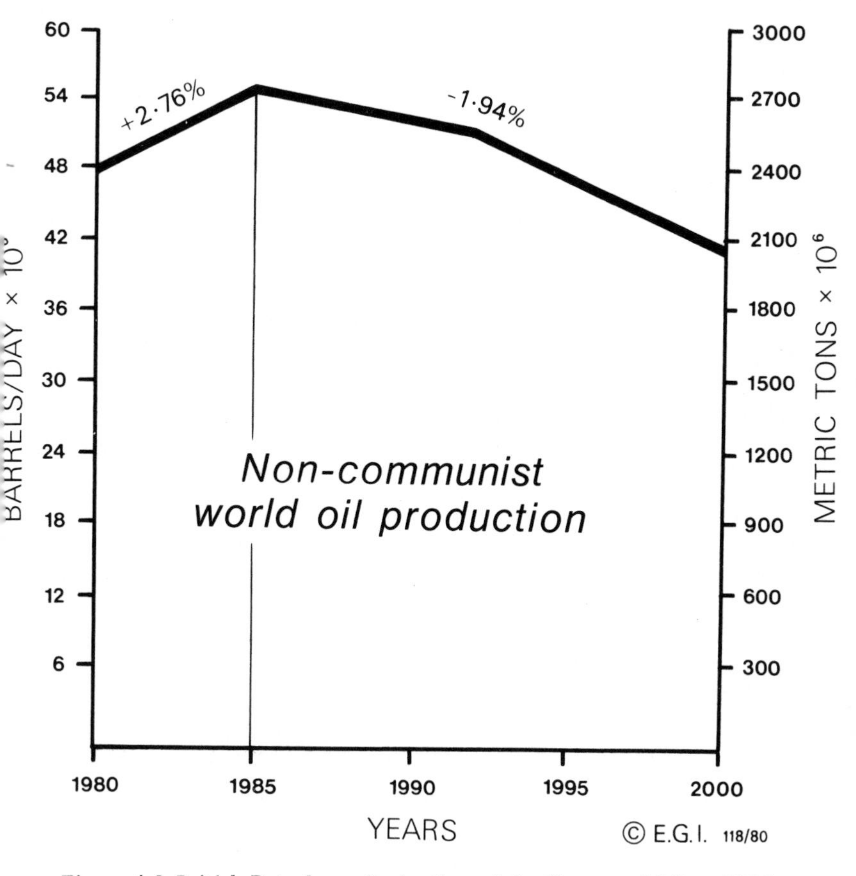

Figure 4.3 *British Petroleum Projection of the Future of Oil to 2000*

BP's Policy Review Unit did not publish its forecasts for the future of oil beyond the year 2000, though a continued decline in the supply of oil thereafter can be inferred from the tenor of the Report. The Report is concerned only with the prospects for the future supply of oil in the non-communist world on the assumption, presumably, that the conditions in and the interests of the communist world as far as oil is concerned can remain more or less isolated from the oil future of the non-communist world. The validity of such an assumption is, however, highly questionable. Both the Soviet Union and China show some interest in exporting oil and/or gas while, in some quarters, it is believed that the Soviet Union, together with Eastern Europe, will become dependent on imports of oil from the rest of the world (= the non-communist world) within five years.[1]

These important policy questions aside, there is the overriding need to make the presentation of an aggregated, long-term future of oil geographically comprehensive. This means using world level data. And, in order to test the implications of the BP view of the future of oil for the size of the world oil resource base, it is necessary to add in a component for the use of oil in the communist world. This has been done from the base of 1980. To coincide with the break-points in the BP non-communist world demand curve, we have added estimates of communist world oil use in 1985 and the year 2000 to the BP non-communist world data for these years. Our estimates are based on the assumption that there will be continued growth in oil use in the communist world for the whole of the period from 1979-2000. For 1980-85, an average annual growth rate of 5.14 per cent is assumed and this has the effect of making the world growth rate somewhat higher than the rate which BP used for the non-communist world, for these years, *viz* 3.4 per cent instead of 2.76 per cent. Post-1985 our assumption is for a communist world average annual rate of growth of oil use of 1.85 per cent. This growth in the communist world partly compensates for BP's assumed fall of 1.94 per cent per annum in non-communist world use from 1985-2000 so that the overall world decline figure now becomes 0.67 per cent per annum. The calculations are set out in Table 4.1.

1. See, for example, the view of the US Central Intelligence Agency. In its most recent statement the CIA projects the communist countries as a group 'to shift from a net export position of 800,000 b/d in 1979 to a net import position of at least one million b/d in 1985'. From the text of Admiral S.Turner's (Director of the CIA) presentation to the *Senate Committee on Energy*, April 23, 1980.

Table 4.1 *Adjustment of BP's estimates of non-communist world oil production, 1980-2000 to a total world basis*

Year	Non-Communist World*		Communist World†		Total World**	
	Production (b/d) x 10^9	*Annual Average Growth Rate*	*Production (b/d) x* 10^9	*Annual Average Growth Rate*	*Production (b/d) x* 10^9	*Annual Average Growth Rate*
1980	48		14.9		62.9	
		+2.76%		+5.14%		+3.4 %
1985	55		19.1		74.1	
		−1.94%		+1.85%		−0.67%
2000	41		25.2		66.2	

* Derived from British Petroleum, *Oil Crisis . . . Again?*, London, 1979, Figure 20.
† Authors' estimates.
** Calculated from * and †.

Using these adjusted figures as the input data, the implications of BP's views on the future of oil are presented in the runs of the model which are illustrated in Figures 4.4 to 4.8. The production curve up to the year 2000 is identical, of course, in all these iterations. For this period it is determined by the defined rates of growth and decline in demand. For the period after the year 2000 the 1985-2000 average annual rate of decline only continues for as long as varying resource base considerations allow. The impact of the constraints will vary according to the defined total future availability of oil and so influence the steepness of the decline curve. In brief, the result of the set of iterations is a test of the potential longevity of the oil industry in respect of variations in the size of the resource base from a minimum of 2000 x 10^9 barrels to a maximum of 11,000 x 10^9 barrels, given BP's declared view that there is a pre-2000 constraint on levels of production.

This BP view of the future means that little new oil needs to be found until the late-1990s because the oil needed until then can be largely secured from the fields already discovered (no matter what the size of the ultimate resource base). With the lowest resource base – of 2000 x 10^9 barrels – the predicated post-1985 slow rate of decline can, as shown in Figure 4.4, continue only until 2003, when production will be somewhat higher than the 1980 level. Immediately thereafter, there would have to be a period of more rapid decline in output as the world's remaining oil – roughly one-third of the total resource

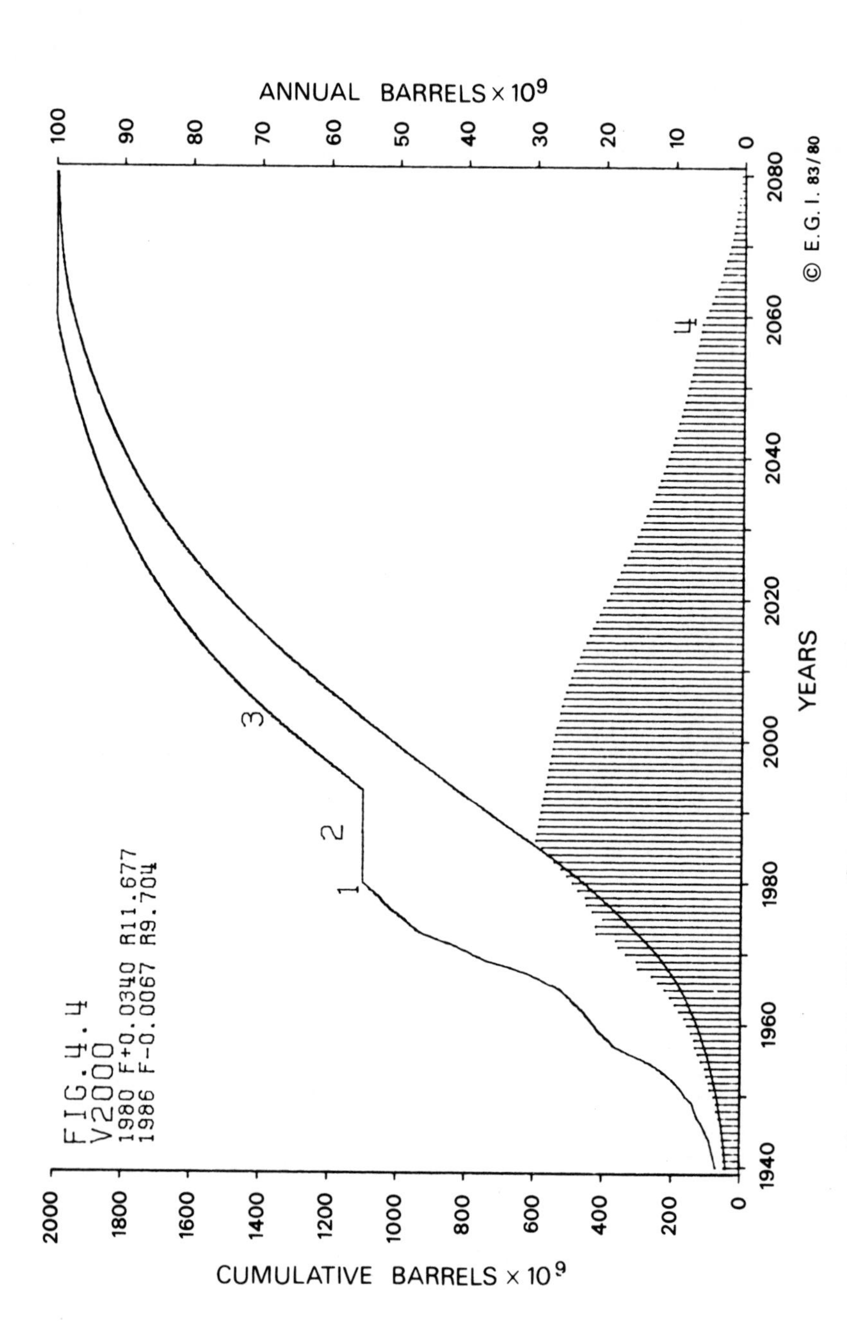
FIG. 4.4
V2000
1980 F+0.0340 R11.677
1986 F-0.0067 R9.704
ANNUAL BARRELS × 10⁹
100
90
80
70
60
50
40
30
20
10
0
CUMULATIVE BARRELS × 10⁹
2000
1800
1600
1400
1200
1000
800
600
400
200
0
YEARS
1940
1960
1980
2000
2020
2040
2060
2080
1
2
3
4
© E.G.I. 83/80

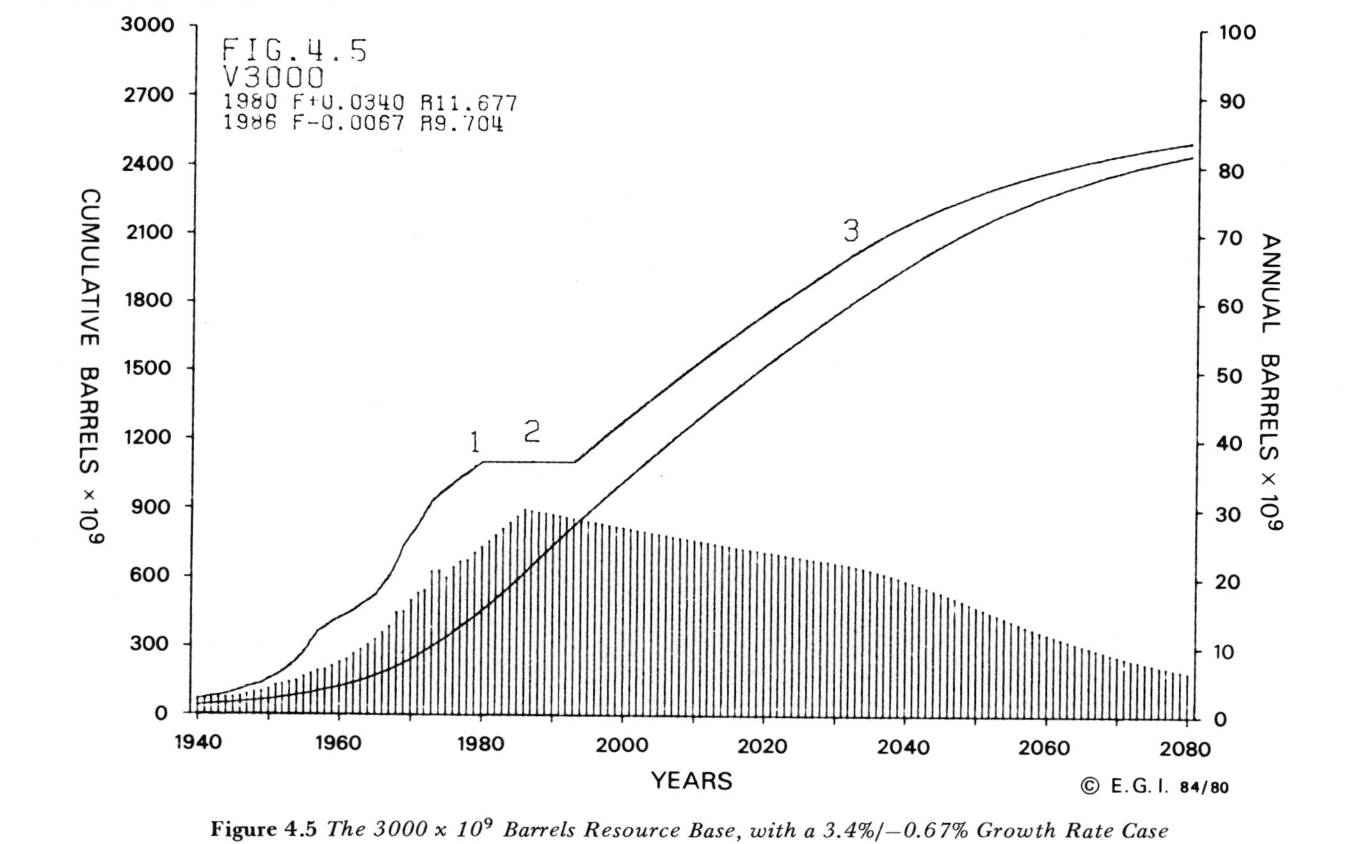

Figure 4.5 *The 3000 x 10^9 Barrels Resource Base, with a 3.4%/−0.67% Growth Rate Case*

base – is more or less used up in the 77 years to 2080. Note, however, that even in this pessimistic case it is the year 2060 before the last of the world's oil is found.

With 3000 x 10^9 barrels (Figure 4.5) the nominated decline curve for the period 1985-2000 can continue through until around 2031, when output is at approximately the same level as that of 1979, before the limited remaining availability of oil requires the decline curve slope to become steeper. Thus, even with this relatively modestly sized resource base the rapid decline of the industry is not necessary until well into the second quarter of the 21st century. Moreover, in this iteration only about 85 per cent of the world's oil has been 'found' by 2080 when production is still at the level of almost 6.4 x 10^9 barrels in the year – greater, that is, than in any year in the history of the industry prior to 1957.

Figure 4.6 shows the results from using a 5000 x 10^9 barrels resource base. In this case the –0.67 per cent decline curve can continue to operate from 1985 through to 2080. By the latter date, which marks the end of the period covered by the simulation, only 54 per cent of the total oil resources of 5000 x 10^9 barrels have been discovered, thus indicating that the gentle decline of the industry could go on well into the 22nd century. With 8000 x 10^9 barrels (Figure 4.7) and 11,000 x 10^9 barrels (Figure 4.8) the picture is basically the same. The decline curves, as defined by the adjusted BP figures, do not, by 1980, generate the need to use more than 35 per cent and 26 per cent, respectively, of the total oil resource bases.

Two further general points may be made concerning all these iterations based on the BP approach. First, currently declared remaining proven reserves of oil (amounting to approximately 650 x 10^9 barrels) are not used up until the year 2003, thus giving much more than adequate or necessary flexibility in terms of the time needed to search for new oil. Second, it is unnecessary in the future to add more than 33.2 x 10^9 barrels to reserves in order to sustain the curve of production predicated by BP. This maximum addition to reserves requirement for the future is one which is little more than the average annual volume of additions to reserves achieved by the industry over the 30-year period since 1950. It is, indeed, a volume of reserves additions which can be achieved, in theory, even if the industry did not find any new oilfields for at least a 20-year period. During this time a one per cent per annum addition to the average rate of recovery of oil from known fields will enable

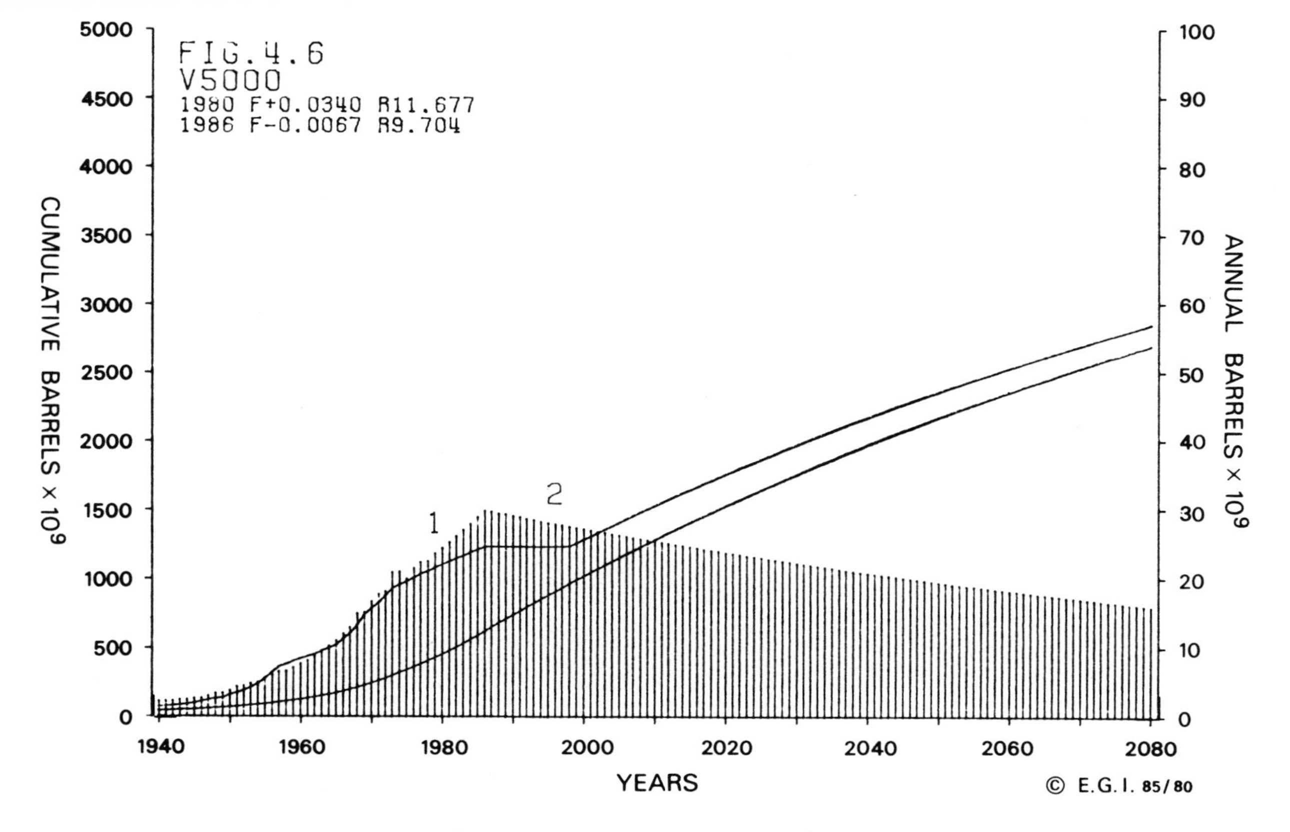

Figure 4.6 *The 5000 x 10⁹ Barrels Resource Base, with a 3.4%/−0.67% Growth Rate Case*

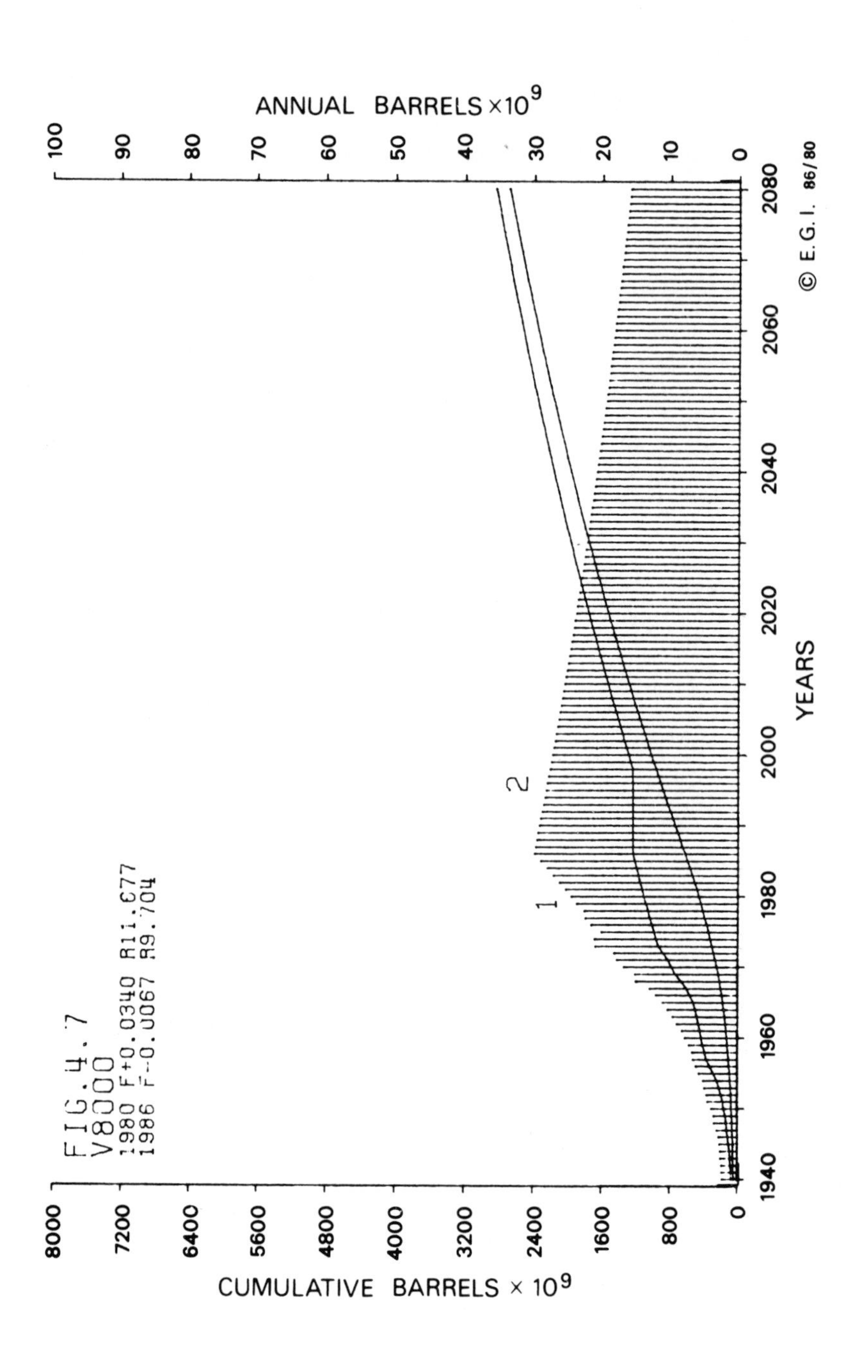

ANNUAL BARRELS ×10⁹
100
90
80
70
60
50
40
30
20
10
0
FIG.4.7
V8000
1980 F+0.0340 R11.677
1986 F-0.0067 R9.704
1
2
1940
1960
1980
2000
2020
2040
2060
2080
YEARS
8000
7200
6400
5600
4800
4000
3200
2400
1600
800
0
CUMULATIVE BARRELS × 10⁹
© E.G.I. 86/80

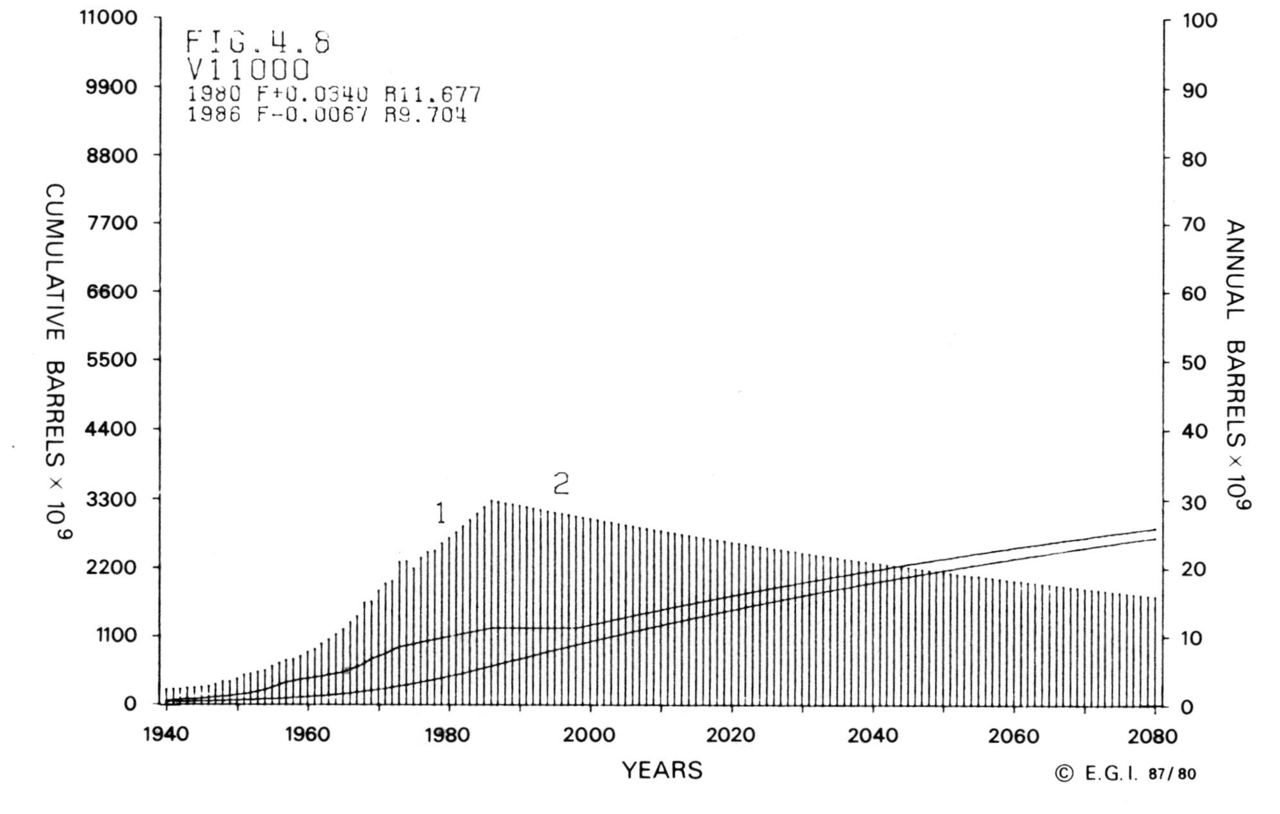

Figure 4.8 *The 11,000 x 10⁹ Barrels Resource Base, with a 3.4%/−0.67% Growth Rate Case*

rather more than this volume of oil to be added to reserves each year.

Overall, the demands in the future which need to be made on the oil industry in relation to the BP's 'Think Tank' expectations are modest in the extreme.[1] It could indeed almost be described as the option which demands no more of the industry than it has achieved to date. It must, therefore, be presented as the most conservative view of the future of oil.

However, in that BP's 'Think Tank' did not take its presentation on the future of oil beyond the year 2000, our inference that the post-1985 decline in use continues unchanged into the 21st century can, of course, be challenged. Instead, it is perhaps the company's expectation that the annual supply of oil will thereafter be held, for as long as possible, at the level achieved in the year 2000. This possibility can be tested through our model. It is also of interest to test the relationship between resources and the required supply in circumstances in which oil output, at some date in the future, falls back to the level of 1979, and is then held constant at that level for as long as the availability of oil makes it possible. These alternative views of the post-2000 development of the industry are illustrated in the two following sets of diagrams.

Figures 4.9 to 4.13 illustrate the maintenance of a year 2000 output plateau for the five resource bases covering the range from 2000 to 11,000 x 10^9 barrels. The plateau rate of output is 27.14 x 10^9 barrels compared with the production of 23.6 x 10^9 barrels in 1979 – an increase of 14.5 per cent. The significant variable in these iterations is, of course, the length of the period over which the year-2000 plateau rate could be maintained. In Figure 4.9 (resource base 2000 x 10^9 barrels) plateau production can be held until 2007; in Figure 4.10 (3000 x 10^9 barrels) until 2026; in Figure 4.11 (5000 x 10^9 barrels) until 2075; and in Figures 4.12 and 4.13 (showing the cases 8000 and 11,000 x 10^9 barrels of resources, respectively), the period of plateau production extends beyond 2080, the end of the study period.

Figures 4.14 to 4.18 present the iterations of the model in which the decline curve is arrested in the year in which the level

1. Had the 1979-85 growth figures and the post-1985 decline figure for future oil production of the 'Think Tank' not been adjusted to take account of expected continued growth throughout the period in the oil industry of the communist world, then the demands on the industry for the non-communist world would have been even more modest than shown here.

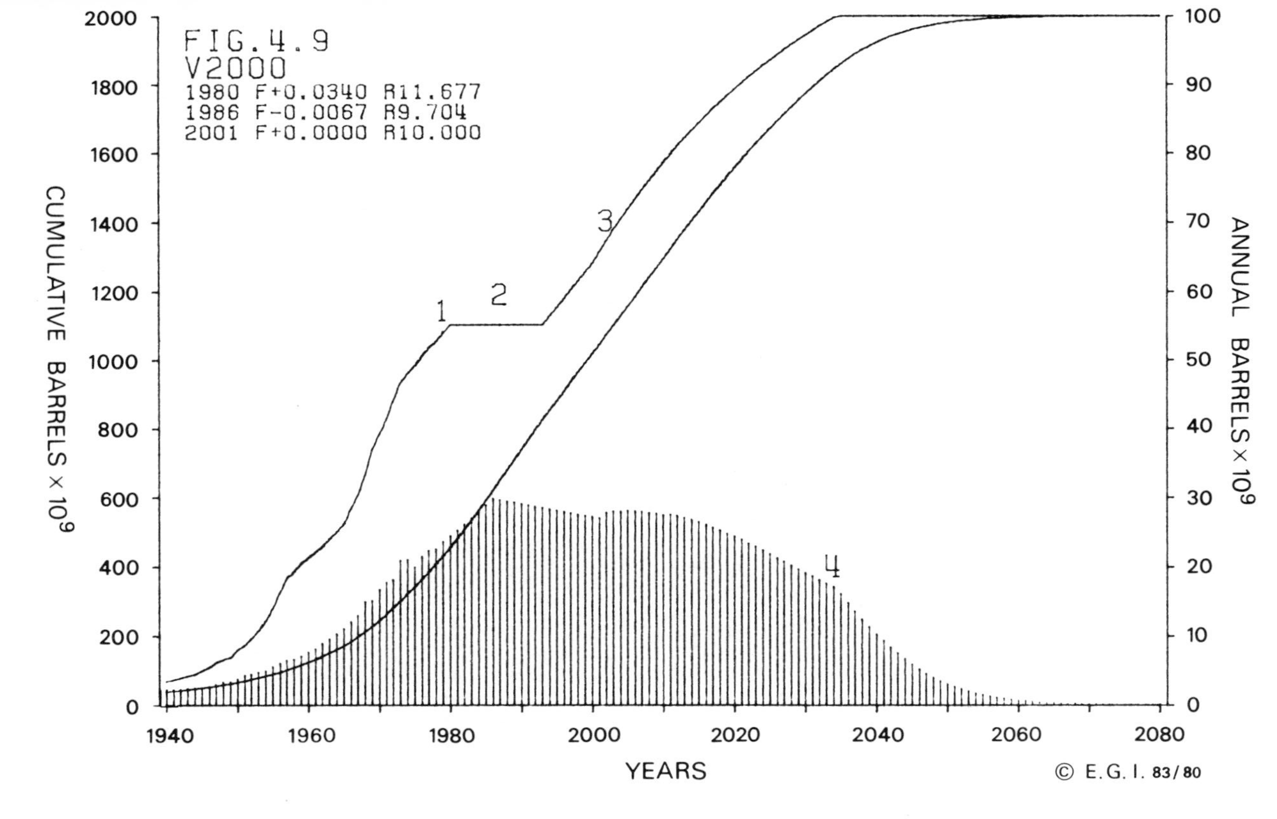

Figure 4.9 *The 2000 x 10^9 Barrels Resource Base, with a 3.4%/−0.67% Growth Rate Case Year 2000 Level Production Plateau*

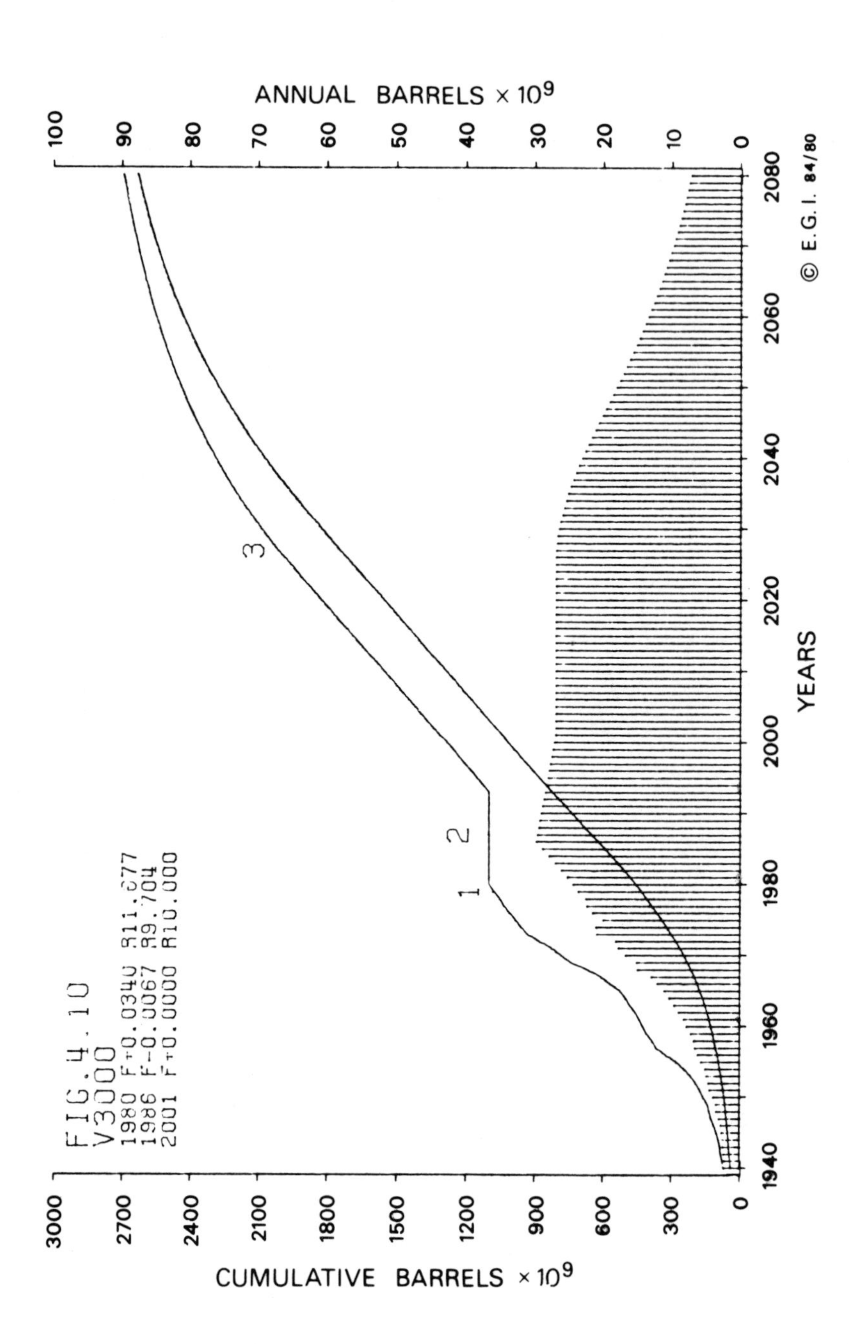
FIG.4.10
V3000
1980 F+0.0340 R11.677
1986 F-0.0067 R9.704
2001 F+0.0000 R10.000
ANNUAL BARRELS × 10^9
100 90 80 70 60 50 40 30 20 10 0
CUMULATIVE BARRELS × 10^9
3000 2700 2400 2100 1800 1500 1200 900 600 300 0
YEARS
1940 1960 1980 2000 2020 2040 2060 2080
1
2
3
© E.G.I. 84/80

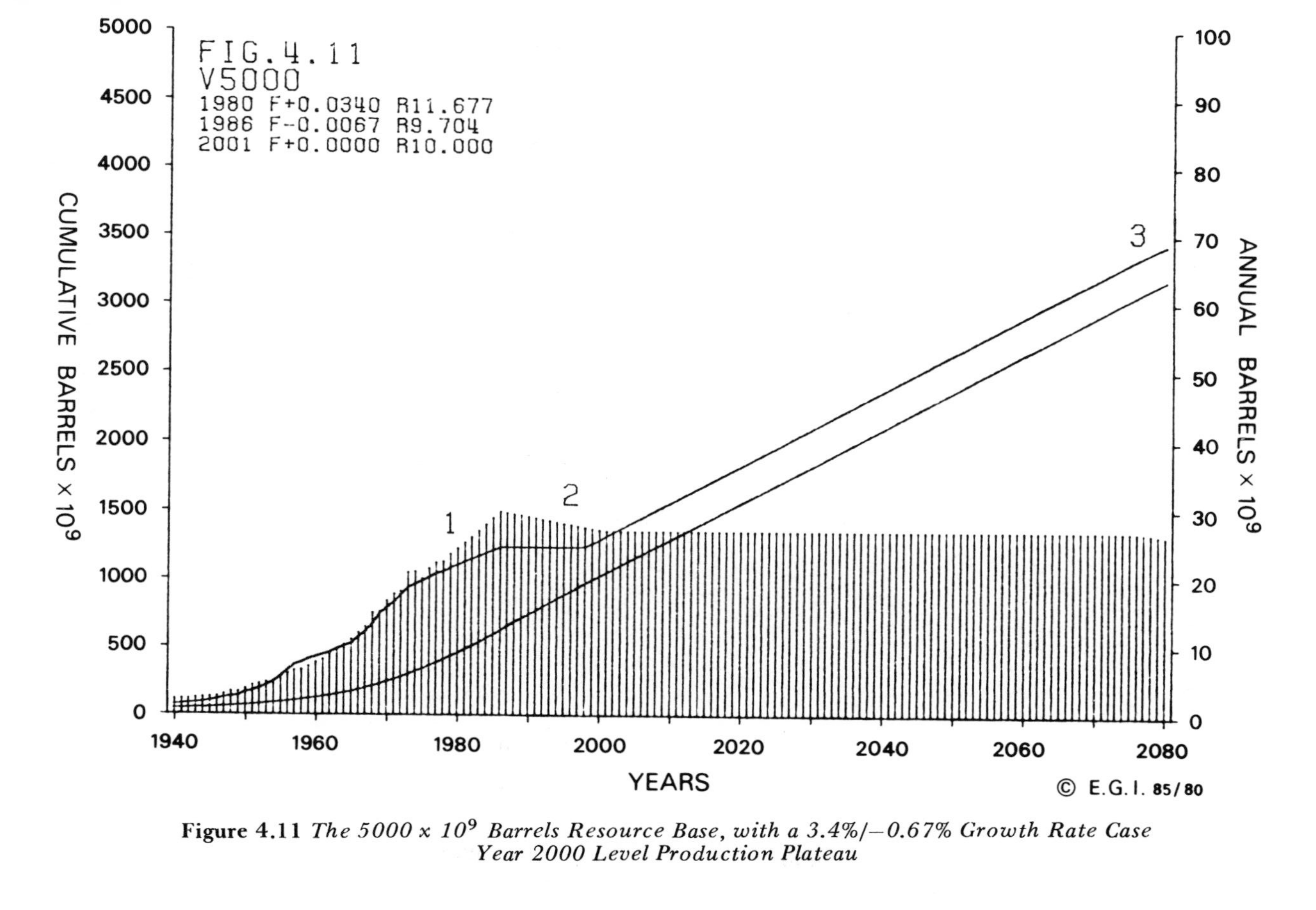

Figure 4.11 *The 5000 x 10⁹ Barrels Resource Base, with a 3.4%/−0.67% Growth Rate Case Year 2000 Level Production Plateau*

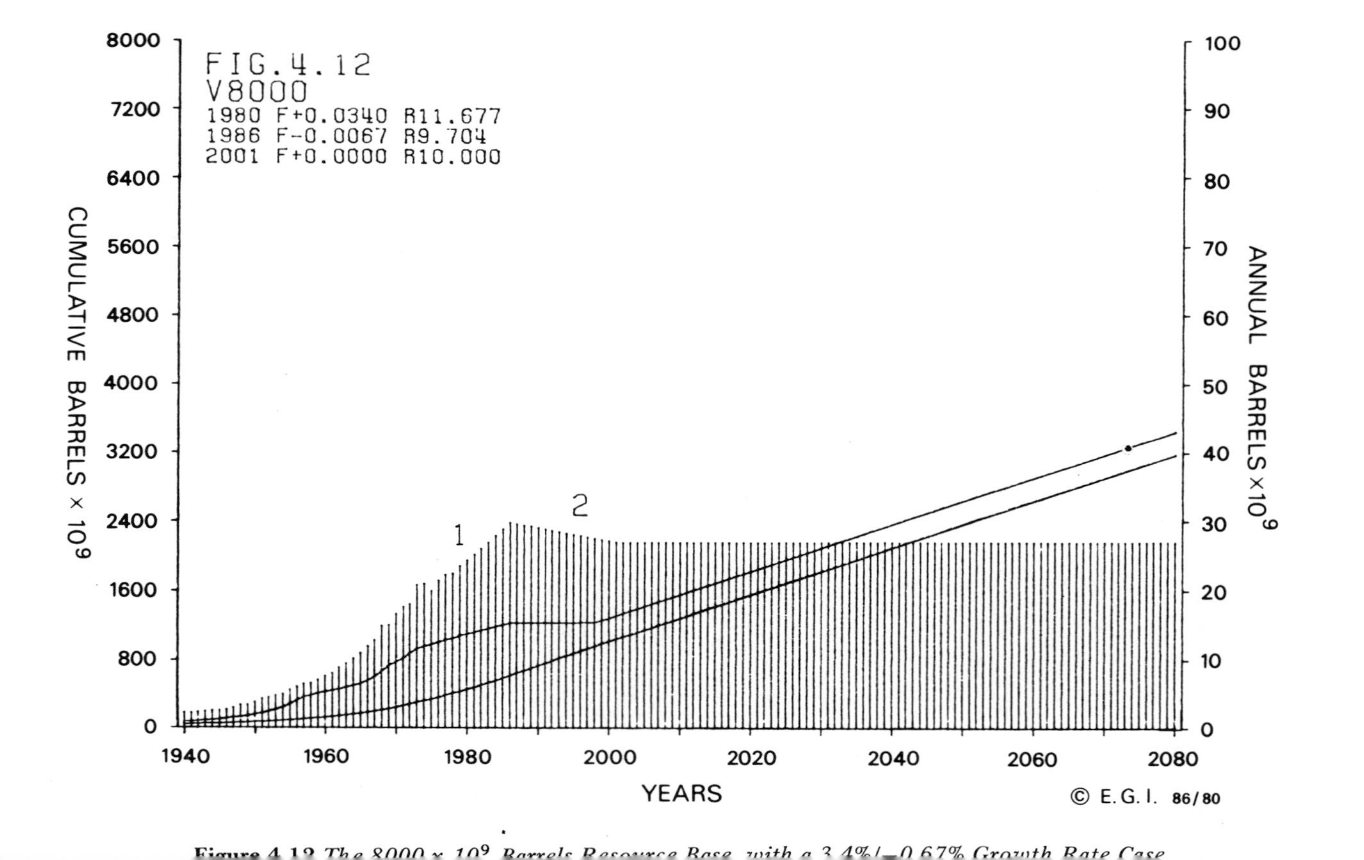

Figure 4.12 *The 8000 × 10^9 Barrels Resource Base with a 3.4%/–0.67% Growth Rate Case*

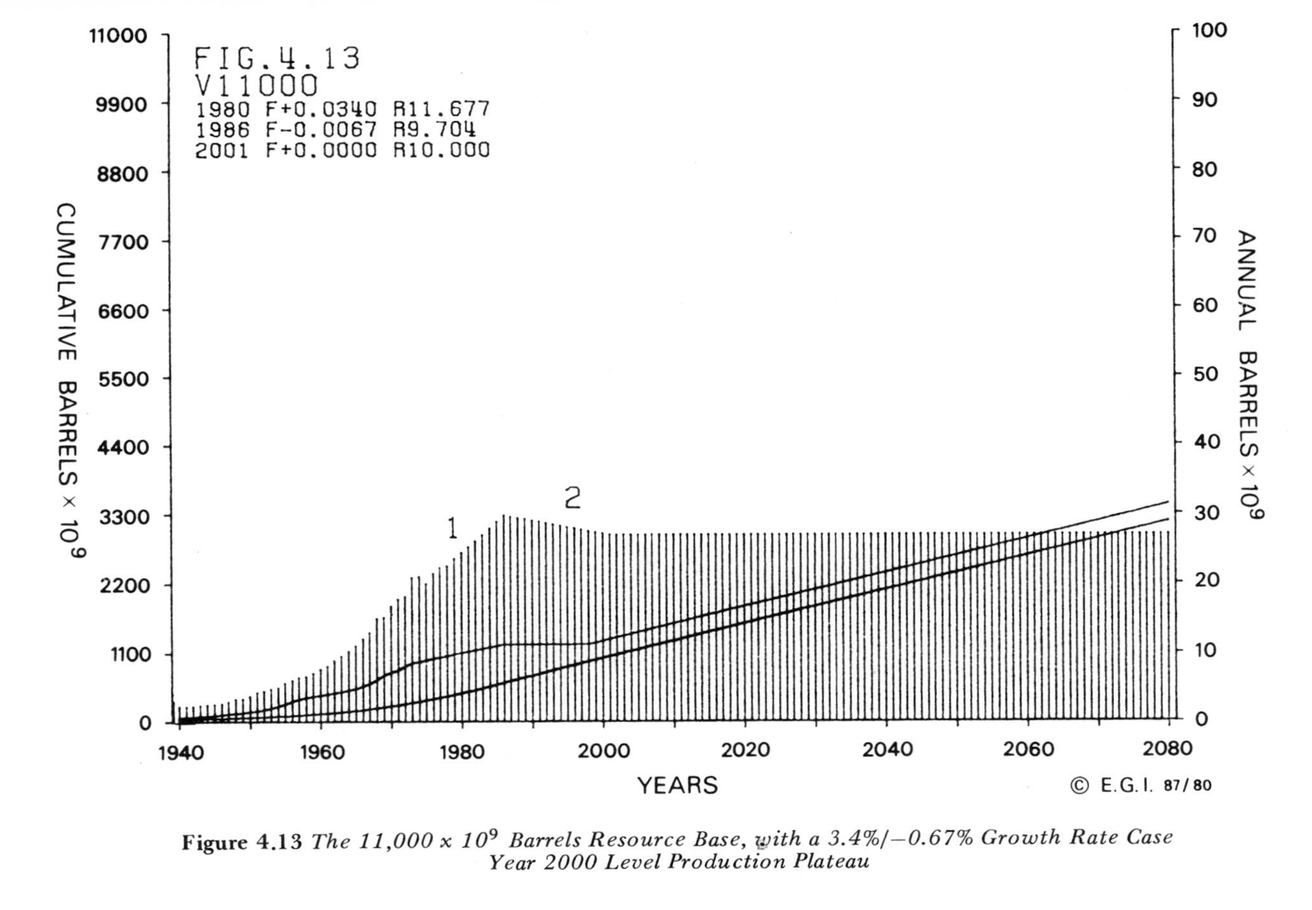

Figure 4.13 *The 11,000 x 10^9 Barrels Resource Base, with a 3.4%/−0.67% Growth Rate Case Year 2000 Level Production Plateau*

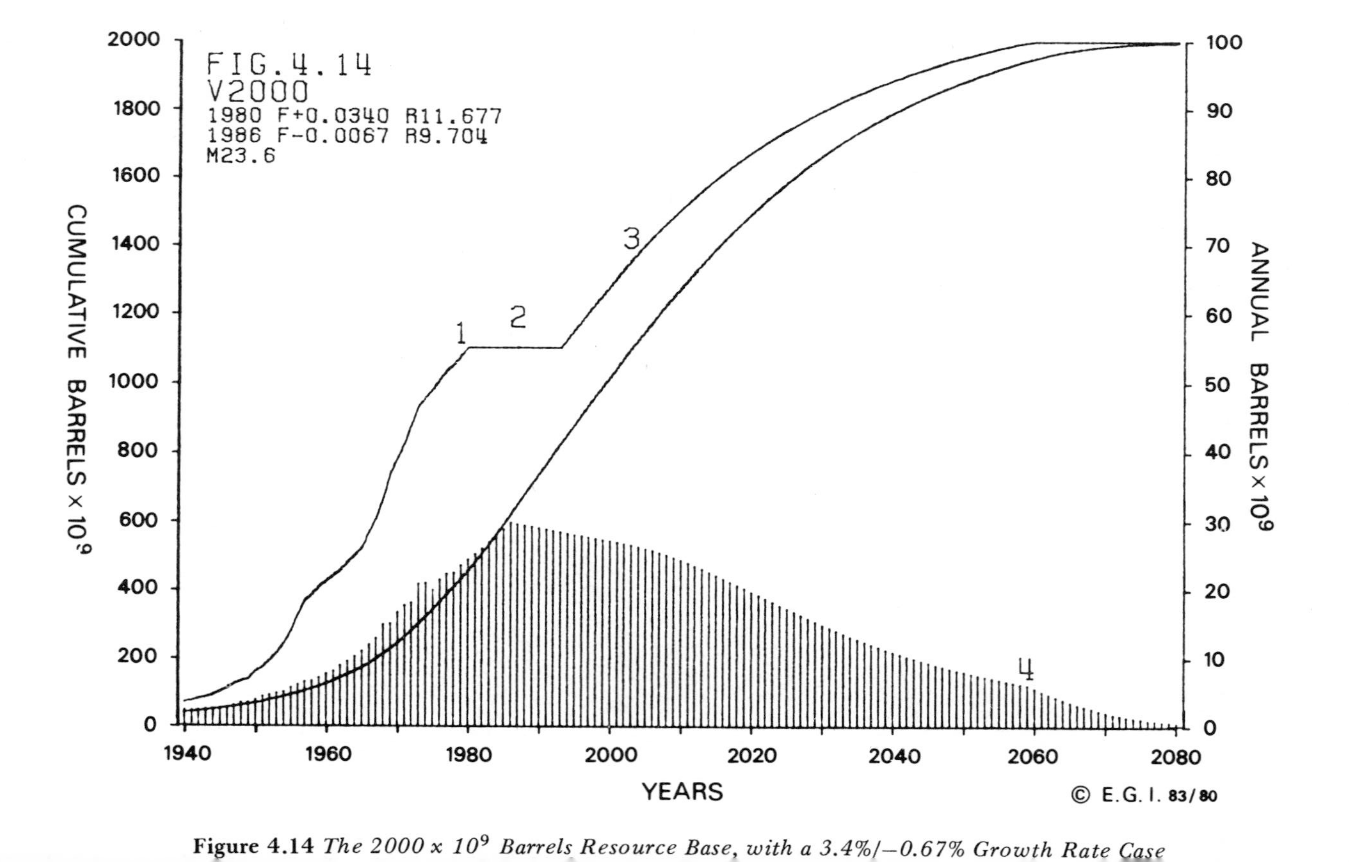

Figure 4.14 *The* 2000×10^9 *Barrels Resource Base, with a 3.4%/−0.67% Growth Rate Case*

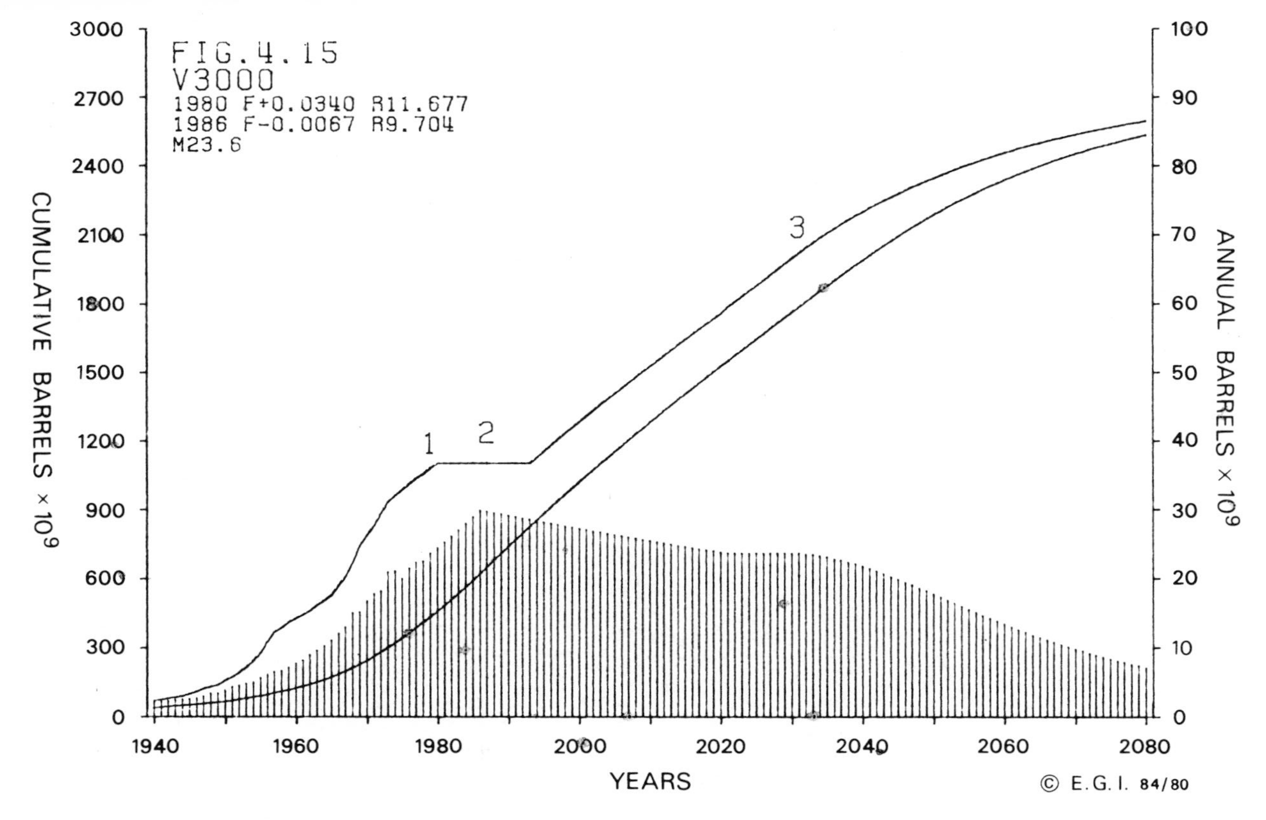

Figure 4.15 *The 3000 x 10^9 Barrels Resource Base, with a 3.4%/−0.67% Growth Rate Case 23.6 x 10^9 Barrels Minimum Production*

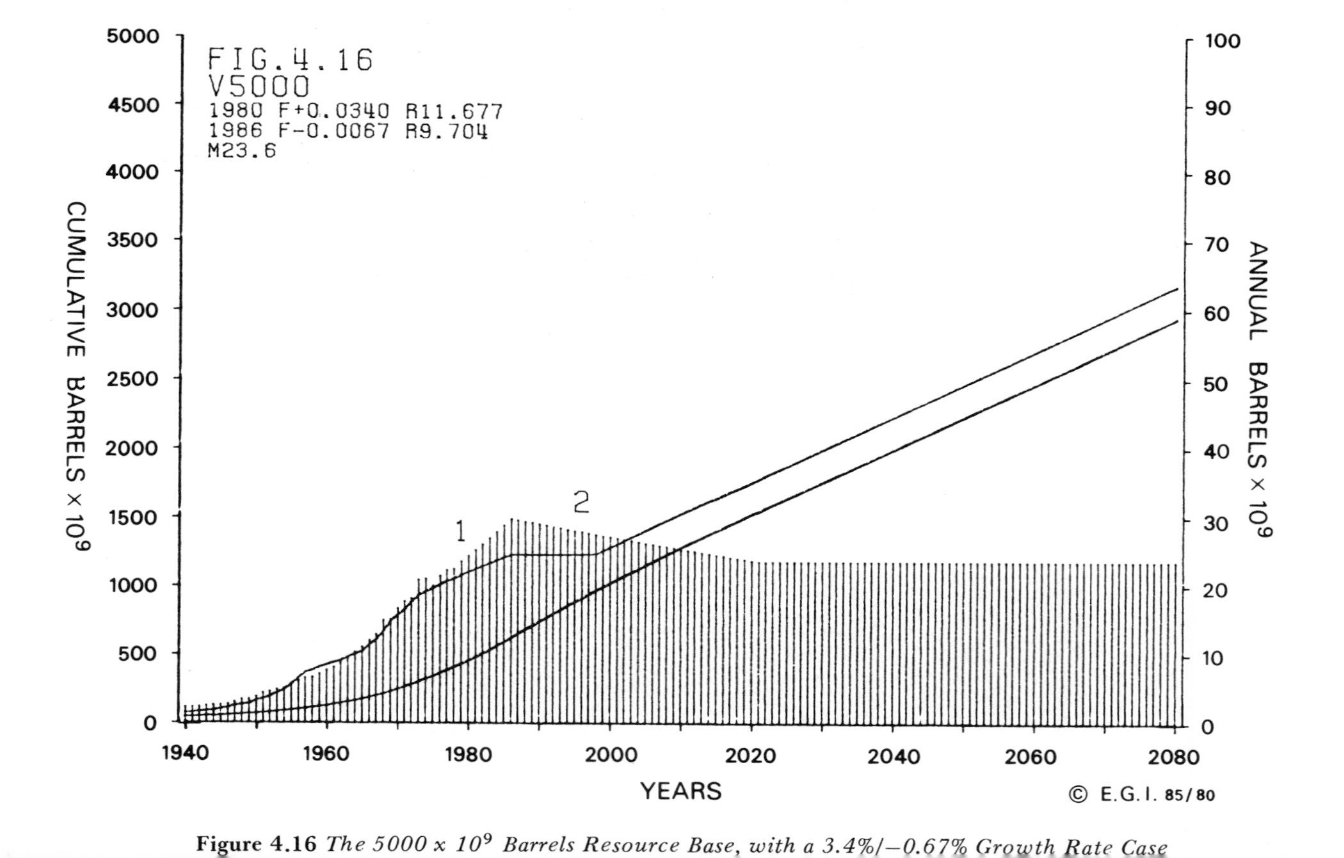

Figure 4.16 *The 5000 x 10^9 Barrels Resource Base, with a 3.4%/−0.67% Growth Rate Case*

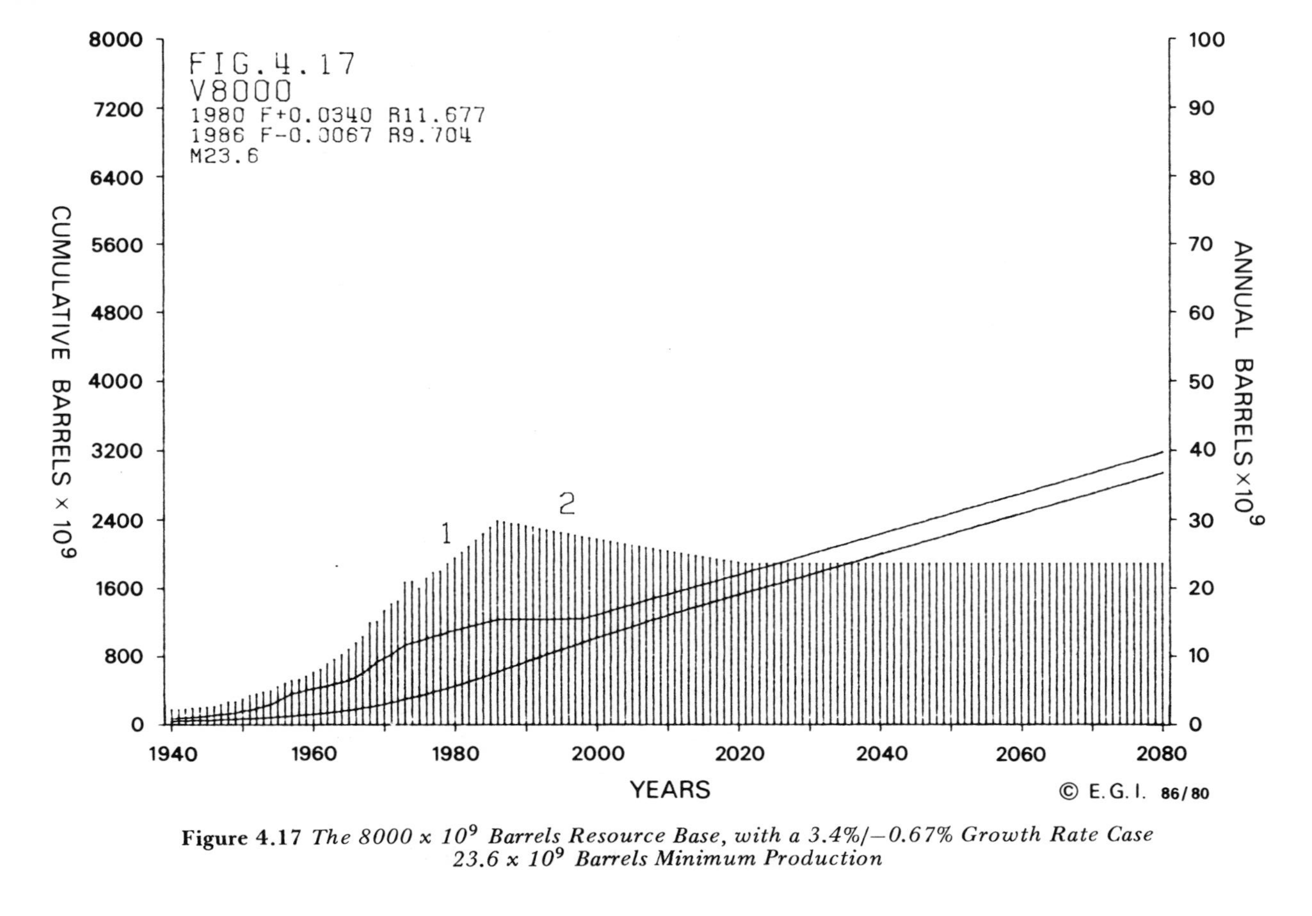

Figure 4.17 *The 8000 x 10^9 Barrels Resource Base, with a 3.4%/−0.67% Growth Rate Case 23.6 x 10^9 Barrels Minimum Production*

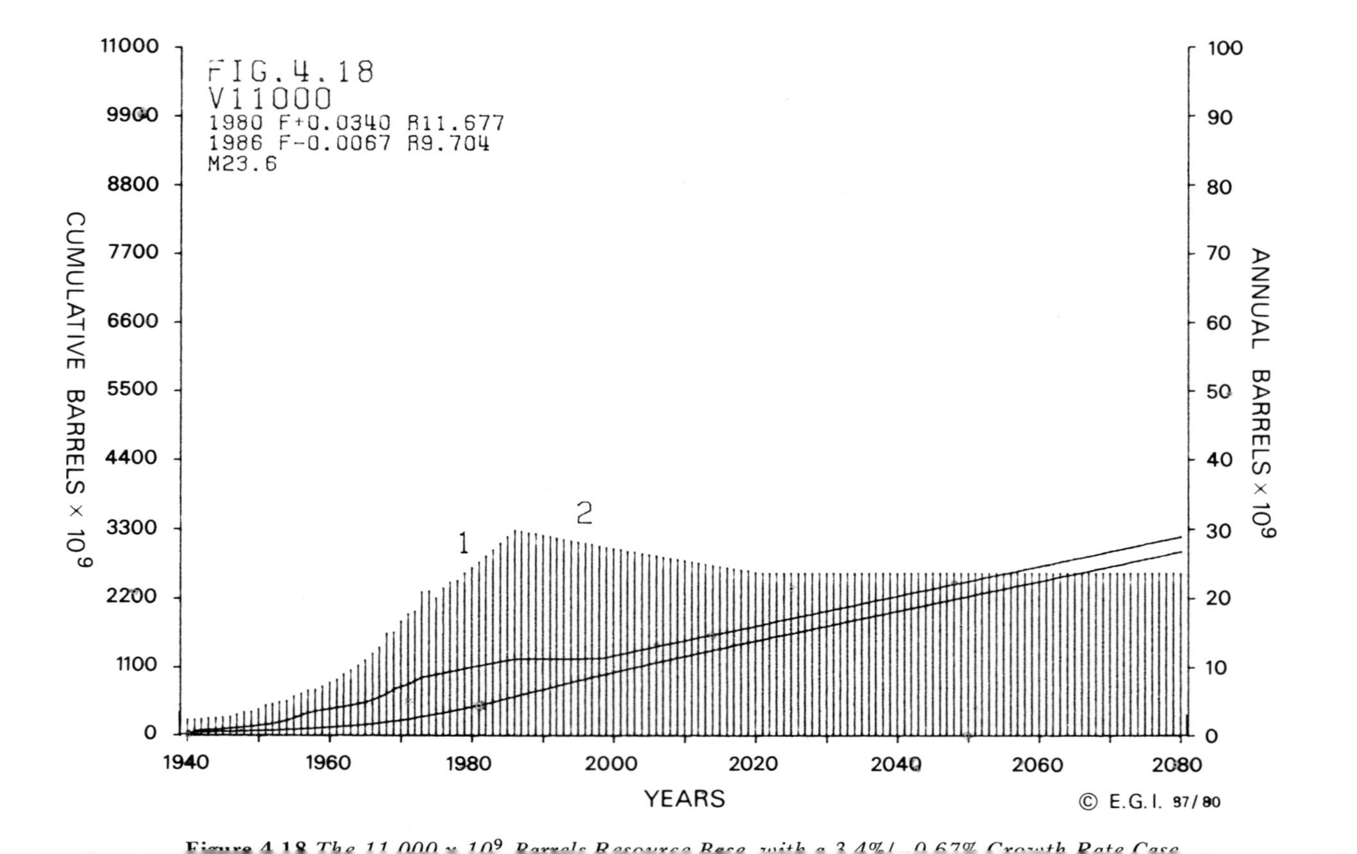

Figure 4.18 *The 11,000 × 10^9 Barrels Resource Base, with a 3.4%/ 0.67% Growth Rate Case*

of production returns to that of 1979. Thereafter, the plateau rate of production (23.6 x 10^9 barrels) is held for as long as the resource base makes this possible. As the inflection date is post-2000 this approach has the effect of extending the period before the onset of the resource-based controlled decline, if the resources are sufficient to make this possible. This is not the case with 2000 x 10^9 barrels (Figure 4.14) but for 3000 x 10^9 barrels (Figure 4.15) it is 2031 before the decline curve again has to set in. For 5000, 8000 and 11,000 x 10^9 barrels (Figures 4.16, 4.17 and 4.18), however, the post-2020 situation (when production is back to the level of 1979) is identical. In each case the plateau production can be maintained throughout the rest of the study period: that is, until after 2080.

The results of these two alternative approaches to the post-2000 future of oil, from the bases provided by BP's presentation of its expectation for the industry from 1980-2000, offer no surprises. Even with the smallest likely resource base the long-term availability of sufficient oil to maintain its use in those economic sectors where substitution is technologically difficult or economically unattractive (eg in transportation and in chemicals production) is clear. For the larger resource bases the plateau of the oil production curve stretches so far into the future and implies the use of so small a part of the total oil potential, that the appropriateness of the post-1985 constraints on production which BP's presentation suggests are necessary, must be called into question.

C. Royal Dutch/Shell's *Energy Imperatives for the Coming Decades*

In October 1979 the President of the Royal Dutch Petroleum Company, Mr D. de Bruyne, gave a paper under this title to an international meeting in Amsterdam.[1] This paper included a graphical presentation of the future of oil in the non-communist world. In this Shell presentation the concern is with the period up to the year 2020, as shown in Figure 4.19.[2]

Whilst Shell's view of the future of non-communist world oil

1. This paper has since been published in *Energy — What Now?*, (Eds. K.E.Davis and P.deWit), Bonaventura, Amsterdam, 1979.
2. As with the data for the BP study (see footnote 4 on page 106), the Shell graphic has also had to be used to retrieve the growth/decline rates in oil supply in the non-communist world — in this case for various time periods up to 2020 — as the details and derivation of the figures are not given in the text.

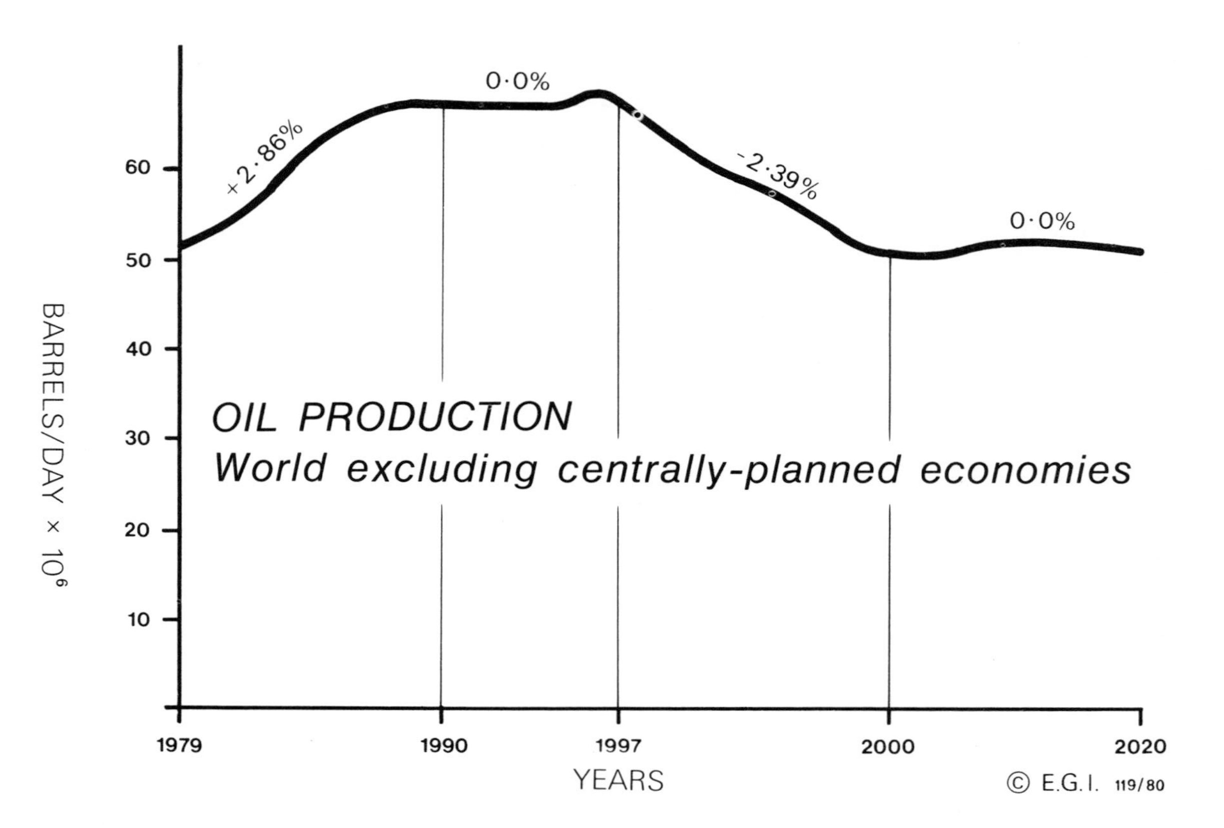

Figure 4.19 Shell Projection of the Future of Oil to 2020

is basically pessimistic in much the same way as that of BP, there are some differences between the presentations of the two companies which are worth mentioning. First, the initial growth period in the Shell study extends to 1990 rather than to 1985, as with BP. During this longer period of growth Shell uses a somewhat higher average annual rate of increase in demand, *viz* 2.86 per cent compared with BP's 2.76 per cent. Thereafter, according to Shell, non-communist world supply will stabilize for seven years before entering into a 12-year period of decline at the relatively sharp rate of 2.39 per cent per annum. Post-2009, however, Shell's graphic shows a more or less stable period of oil production through to 2020, the terminal date of its study. From this indication of stability up to 2020 we shall assume, for the purposes of our analysis, that the supply curve will maintain this shape for as long as the resource base makes it possible after that date.

However, in order to adjust Shell's data for the non-communist world to the global position, data for the communist world has again had to be added. This was done in exactly the same way as in the BP case and the results in this instance are shown in Table 4.2. By adding in the data for the communist world the overall world-wide increase in supply to 1990 now rises to three per cent per annum. Thereafter, there is a modest growth rate of 0.47 per cent per annum for the period 1991-97 and from 1998-2009 the rate of decline is reduced from 2.39 per cent to 1.34 per cent per annum. From 2010 to 2020 the trend reverts to a growth rate of just under one-half per cent. In extrapolating this adjusted Shell data there is, of course, a growth in the supply of oil after 2020, compared with the decline for the post-2000 period in the adjusted BP model. It is only a small growth rate of 0.44 per cent per annum, and it is this rate which is built into the adjusted Shell simulation of world oil supply. In the iterations of the model this continues as long as such growth is possible when related to the assumed size of the resource base.

The five iterations of the adjusted Shell data are presented in Figures 4.20 to 4.24. With the smallest resource base – 2000 x 10^9 barrels – (shown in Figure 4.20), the supply of oil reaches its peak in 1996. This is because two-thirds of all oil has by then been found and the level of production must (as defined in the model) start to fall away. This is, indeed, not a very different outcome from that presented by Shell for the period to the turn of the century – as can be seen by comparing

Table 4.2 *Adjustment of Shell's estimates of non-communist world oil production, 1979-2020 to a total world basis*

Year	Non-Communist World*		Communist World†		Total World**	
	Production (b/d) x 10^9	*Annual Average Growth Rate*	*Production (b/d) x 10^9*	*Annual Average Growth Rate*	*Production (b/d) x 10^9*	*Annual Average Growth Rate*
1979	49.5		14.4		63.9	
		+2.86%		+3.39%		+3.0 %
1990	67.5		20.8		88.3	
		+0.0 %		+1.76%		+0.47%
1997	67.5		23.5		91.0	
		−2.39%		+1.14%		−1.34%
2009	50.5		26.9		77.4	
		+0.0 %		+1.15%		+0.44%
2020	50.5		30.5		81.0	

* Derived from data in D. de Bruyne, 'Energy Imperatives for the Coming Decades', in: *Energy – What Now?*, (K.E.Davis and P. de Wit, Eds.), Amsterdam, 1979.
† Authors' estimates.
** Calculated from * and †.

Figure 4.19 with Figure 4.20. The difference between the output of the model and Shell's view lies in the absence in the former of a post-2009 period of stability in output. Our model indicates the need for a continued decline in oil supply in the early part of the 21st century in order to prevent a much steeper rate of fall off in supply later in the century. At that time, even with the smallest resource base, oil will still be required in considerable quantities for uses which are not substitutable by other sources of energy.

At 3000 x 10^9 barrels (in Figure 4.21) the longer-term outlook is significantly different. With this size resource base the post-2009 up-turn in supply is possible for a decade, so that there is a clearly defined second peak in 2021 with a production in that year of over 30 x 10^9 barrels – at a level which is one-third more than in 1979. Thereafter, there is the slow decline curve in output (as modelled) so that even as late as 2065 over 10 x 10^9 barrels of oil can be produced to provide for essential oil needs. By 2080 almost all the assumed total of reserves (2863 out of 3000 x 10^9 barrels) have been found.

On moving to look at the implications of a 5000 x 10^9 barrels resource base (Figure 4.22) the amount of oil still available after

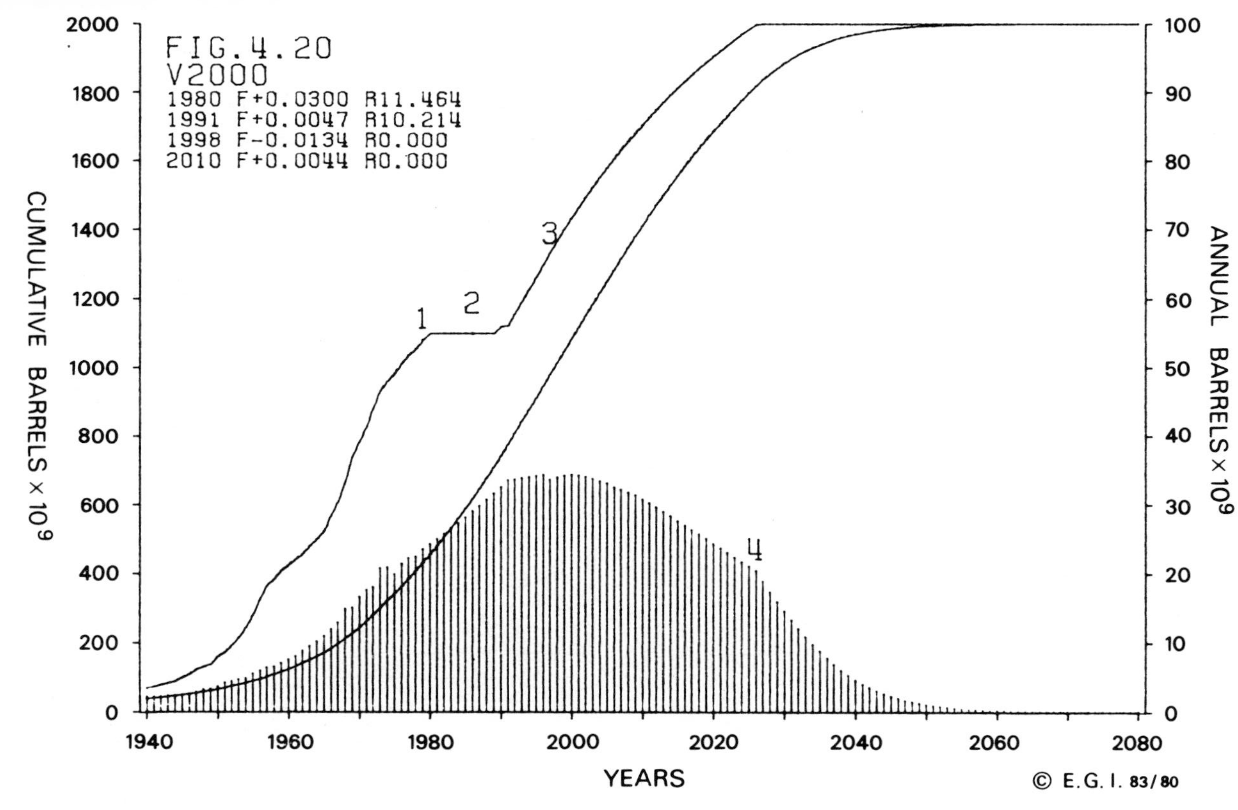

Figure 4.20 *The 2000 x 10^9 Barrels Resource Base, with a 3%/0.47%/−1.34%/0.44% Growth Rate Case*

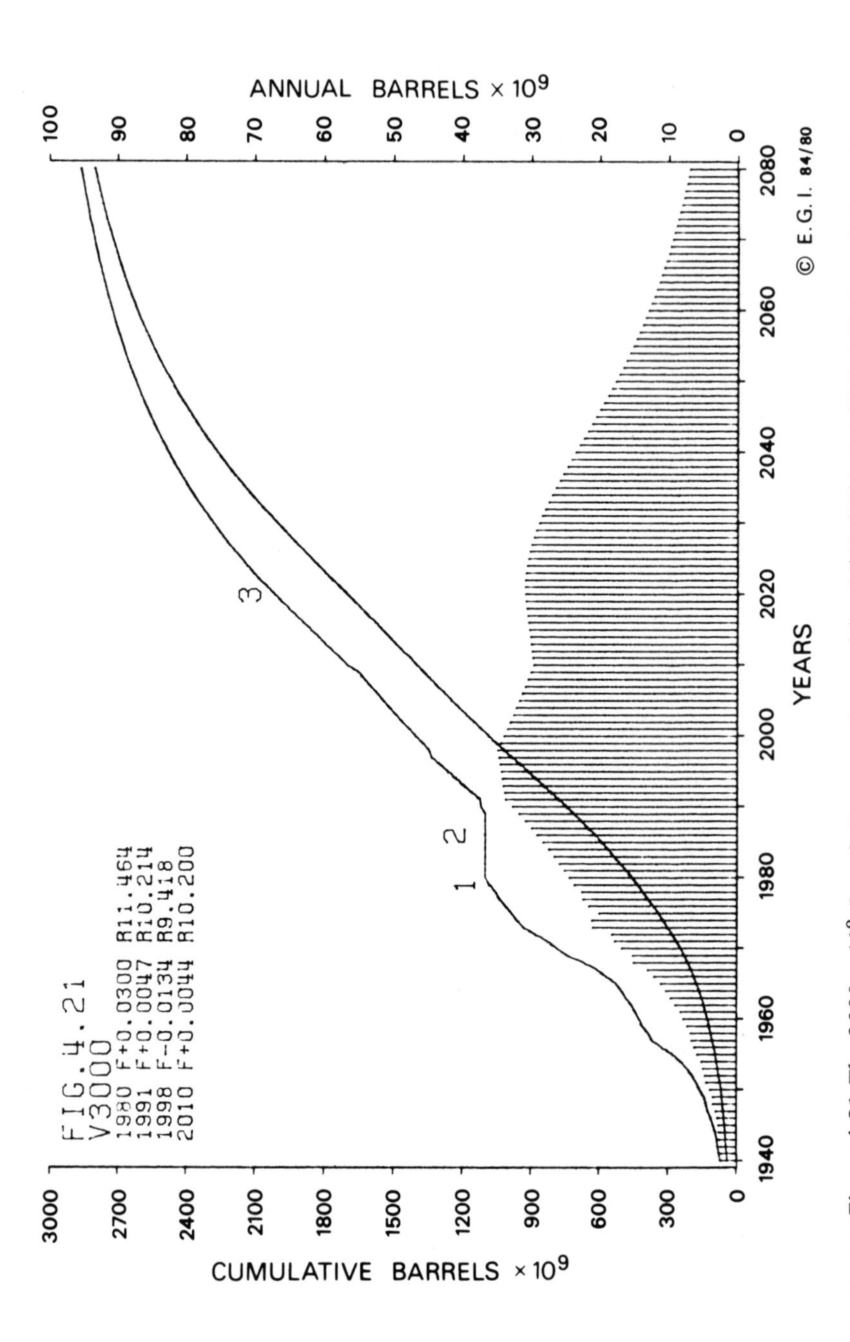
FIG. 4.21
V3000
1980 F+0.0300 R11.464
1991 F+0.0047 R10.214
1998 F-0.0134 R9.418
2010 F+0.0044 R10.200
1
2
3
ANNUAL BARRELS × 10^9
CUMULATIVE BARRELS × 10^9
YEARS
0
10
20
30
40
50
60
70
80
90
100
0
300
600
900
1200
1500
1800
2100
2400
2700
3000
1940
1960
1980
2000
2020
2040
2060
2080
© E.G.I. 84/80

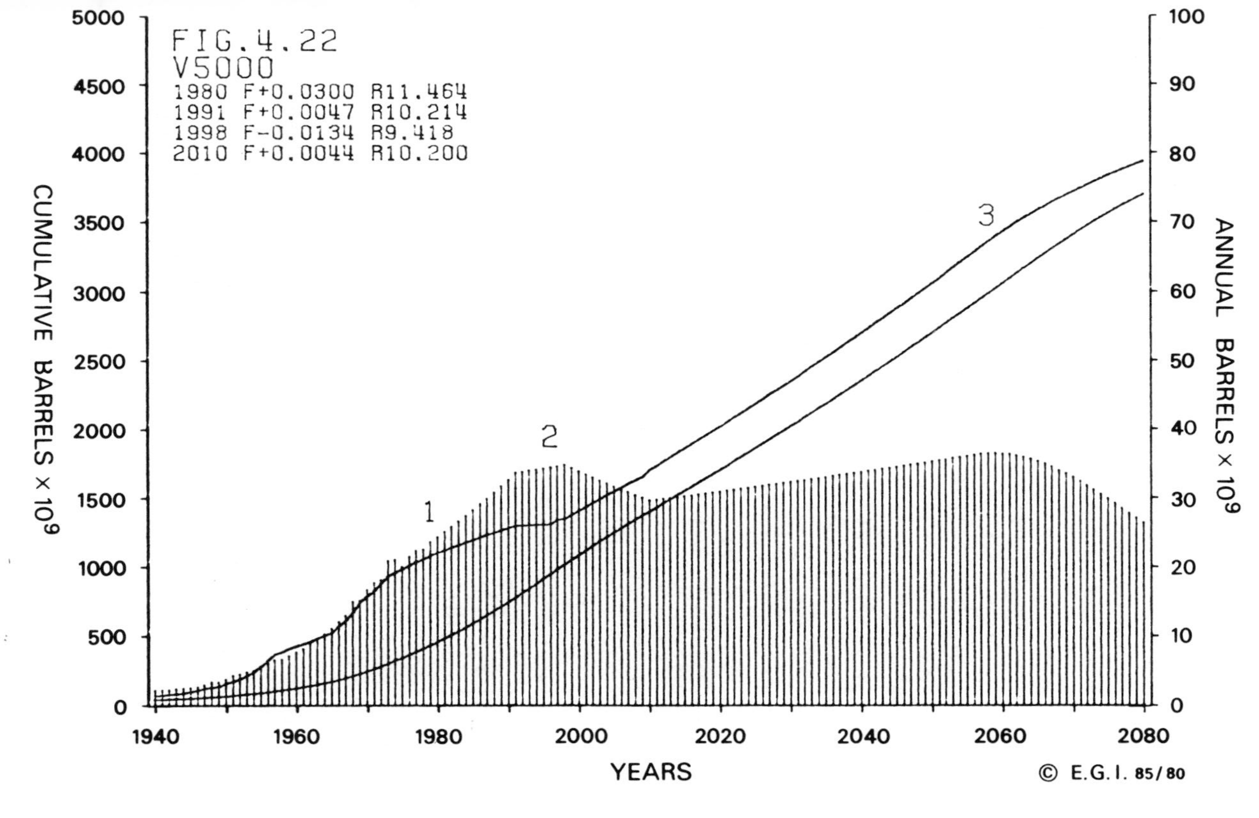

Figure 4.22 *The 5000 x 10^9 Barrels Resource Base, with a 3%/0.47%/−1.34%/0.44% Growth Rate Case*

the initial late 20th century peak (and the subsequent decline in output from 1998) is sufficient to enable the post-2009 upward trend in supply to continue for most of the rest of the study period: *viz* to 2058 when an all-time peak of 36.4 x 10^9 barrels is achieved (compared with the late 20th century peak of just under 35 x 10^9 barrels). By 2080 the requirements of the oil supply pattern have necessitated the discovery of less than 70 per cent of the assumed 5000 x 10^9 barrels resource base.

For both the resource bases of 8000 and 11,000 x 10^9 barrels (Figures 4.23 and 4.24, respectively) the post-2009 growth period in the industry extends beyond the end of the study period in 2080. By the latter date output is over 40 x 10^9 barrels and is still growing at the rate defined in Table 4.2. By that time only just over 50 per cent of the 8000 x 10^9 barrels resource base has had to be discovered and just under 40 per cent of the 11,000 x 10^9 barrels base.

With these latter sized resource bases it is clearly evident that the amounts of available oil do not constitute relevant constraints on the future of oil – not at least for the period up to 2080. Thus oil prospects by no means necessarily create the 'energy imperative' stated by Mr De Bruyne. His imperatives, involving a rapid change to other energy sources, emerge only out of the limited resource base cases of 2000 and, possibly, 3000 x 10^9 barrels. Even in this latter case oil supply continues to expand to the year 2021 – some 41 years hence. In the 5000 x 10^9 barrels case any 'imperative' to change from oil only emerges after the middle of the 21st century when viewed from the point of view of the global oil supply/demand situation. In the meantime, if there are such imperatives they arise from other considerations. These could include the inability of the oil industry, as constituted, to cope with the rapidly changing conditions for the exploitation of oil in many parts of the world. This includes regions where the Western world's existing oil institutions find it difficult to work, or regions from which the international oil companies are excluded. We shall return to essentially politico-economic aspects of the outlook for oil in Chapter 6.

D. 'World Energy Demand to 2020' Case

One very recent influential presentation of the 'energy problem' has been that of the Workshop on Alternative Energy Strategies

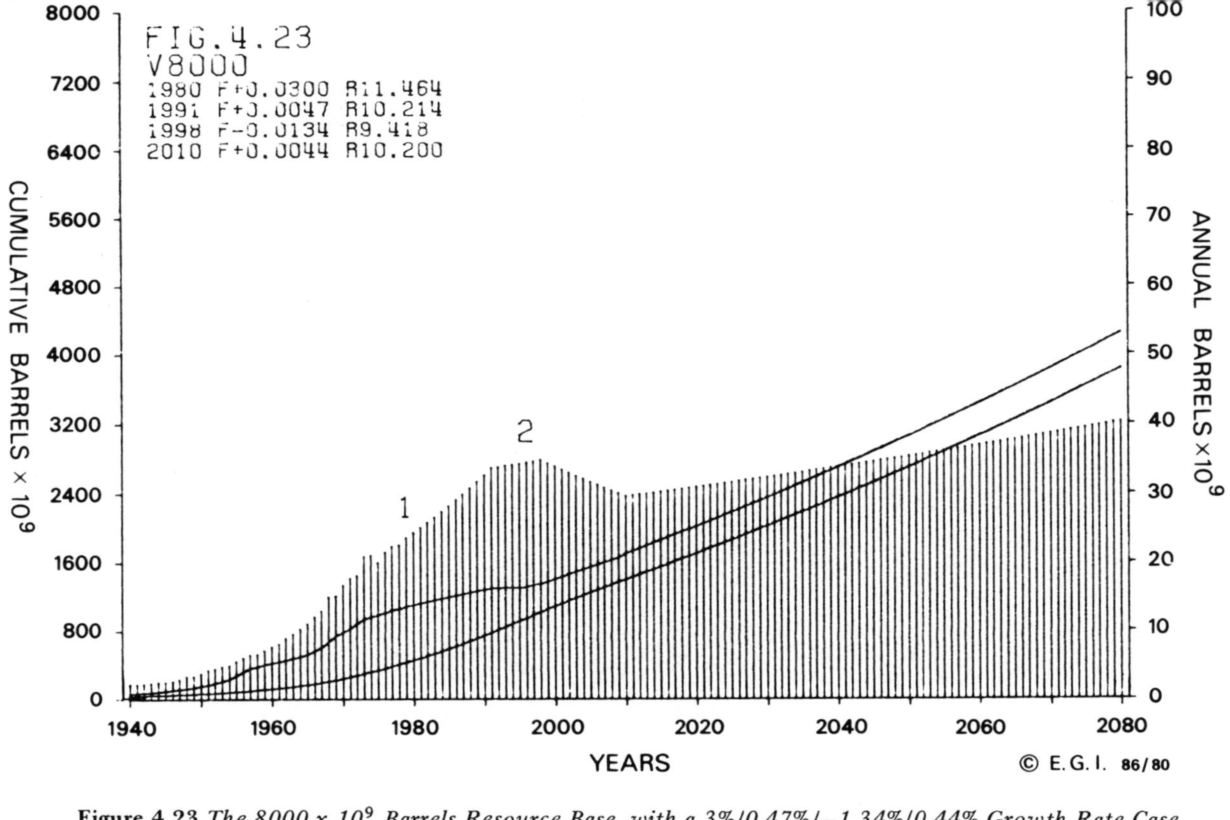

Figure 4.23 *The 8000 x 10^9 Barrels Resource Base, with a 3%/0.47%/−1.34%/0.44% Growth Rate Case*

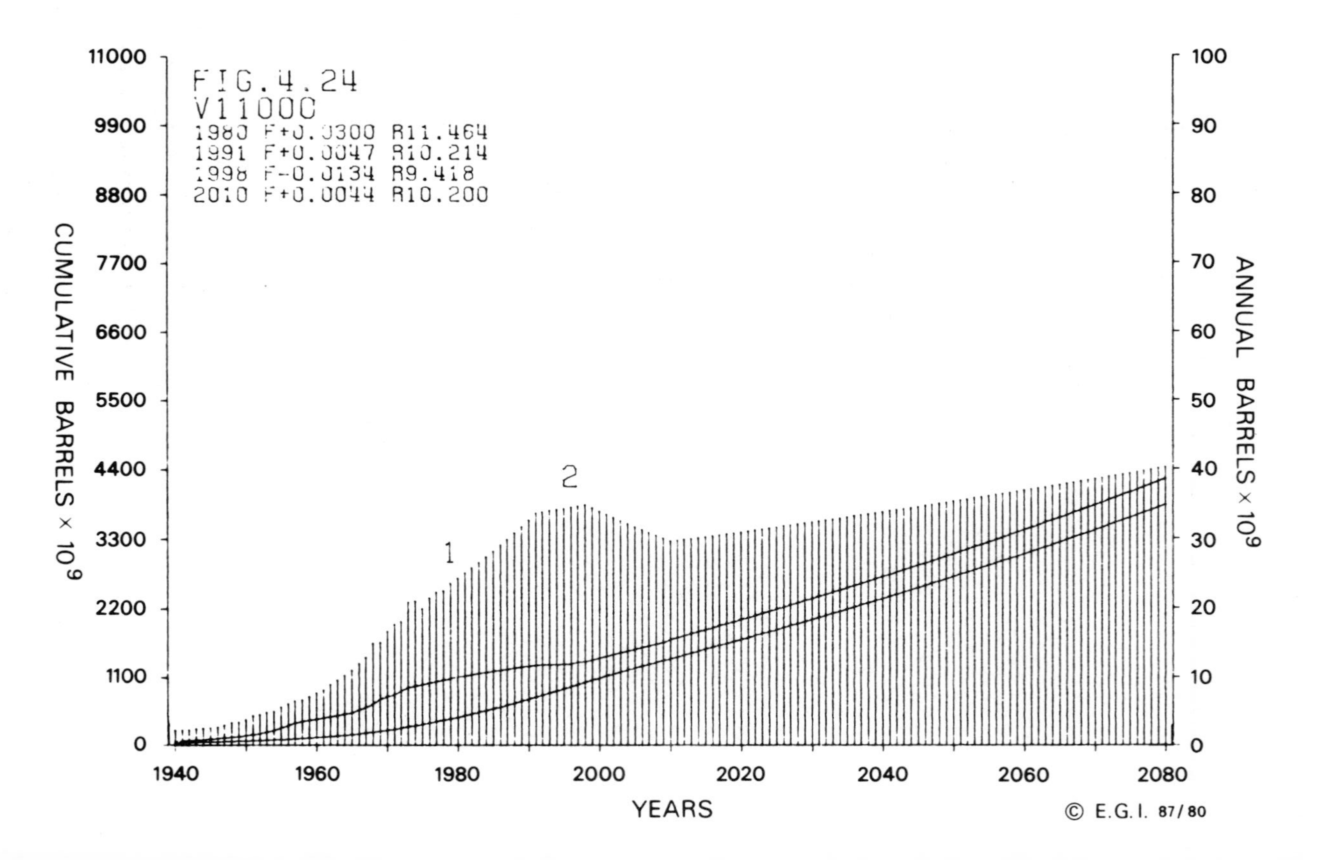
FIG. 4.24
V1100C
1980 F+0.0300 R11.464
1991 F+0.0047 R10.214
1998 F-0.0134 R9.418
2010 F+0.0044 R10.200
1
2
CUMULATIVE BARRELS × 10^9
11000
9900
8800
7700
6600
5500
4400
3300
2200
1100
0
ANNUAL BARRELS × 10^9
100
90
80
70
60
50
40
30
20
10
0
1940
1960
1980
2000
2020
2040
2060
2080
YEARS
© E.G.I. 87/80

(WAES).[1] It gave a generally pessimistic view of the future of the supply of oil – a hardly surprising outcome in the light of the close involvement of six oil companies – including Shell and BP – in its deliberations. Responsibility for the oil supply/demand relationships in the study lay largely with the Energy Research Group (ERG) attached to the Cavendish Laboratories in the University of Cambridge and to which Shell and BP consultants and/or seconded personnel made an important contribution on the oil supply side. This Cambridge Group was also responsible for writing a 'World Energy Demand to 2020' study as part of the World Energy Conference's report for its 1978 annual international conference.[2] Limits on the availability of oil have been a favourite input for this Cambridge Group's work on the future energy outlook. For the purposes of its world energy demand study the Group established ranges for oil supply availability limits for the period up to 2020. Their

Table 4.3 *Energy Research Group, Cavendish Laboratory, University of Cambridge: estimates of world oil supply, 1978-2020**

Year	Low Estimate	High Estimate (*x 10^9 barrels*)	Mid-Point	Average Annual Growth Rate
1978 (*actual*)	22.5	22.5	22.5	
				1.6 %
1985	23.4	26.9	25.15	
				1.15%
2000	26.3	33.4	29.85	
				−0.55%
2020	21.7	31.8	26.75	

* Derived from data in 'World Energy Demand to 2020', World Energy Conference, *World Energy Resources 1985-2020*, IPC Science and Technology Press, Guildford, 1978.

conclusions on the future availabilities of oil are set out in Table 4.3. From these data average growth/decline rates have been established for the periods 1978-85, 1986-2000 and 2001-20. These are also shown in Table 4.3. In the case of this study – unlike those of BP and Shell – the total world position was the base for the presentation. Thus no geographically-related adjustment factors are required in this instance.

1. *Energy: Global Prospects 1985-2000*, Report of the Workshop on Alternative Energy Strategies, McGraw Hill, New York, 1977.
2. World Energy Conference, *World Energy Resources 1985-2020*, IPC Science and Technology Press Ltd., Guildford, 1978.

Figures 4.25 to 4.29 relate this oil component in world energy demand up to 2020 to the five different world oil resource bases. With a 2000 x 10^9 barrels resource base (Figure 4.25) we find a beginning-of-the-21st-century peak in production (peak year 2004) followed by a decline which immediately has to be steeper than the defined rate of –0.55 per cent per annum. This is because it becomes a function of the production constraint built into the model once two-thirds of total reserves have been discovered. In this iteration all the oil is found by 2031. By 2080 it is all but used up and there is an annual production post-2055 of less than 10 x 10^9 barrels per year. In this case the oil industry would, in essence, cease to be important by 2023 (when production would have fallen from its peak in 2004 of 31.3 x 10^9 barrels to a level lower than that of 1979) and it would cease to exist by the beginning of the second half of the 21st century.

This clearly presented view of the currently held conventional view of the future of the oil industry is quite optimistic, compared, that is, with those institutions, such as BP, which see the industry as reaching its peak in the mid-1980s. It is, nevertheless, a conclusion which can be called into question simply by relating the Cambridge Group's interpretation of the future of oil to an ultimate resource base of 3000 x 10^9 barrels. The result of this relationship is shown in Figure 4.26. There is, of course, no change, between this iteration and that for 2000 x 10^9 barrels (see Figure 4.25), in the build up of the industry to the year 2000, as this is a function of the growth rates which have been defined by the Cambridge Group. Thereafter, however, the larger resource base enables the defined decline in production of 0.55 per cent per annum to continue for almost a quarter of a century (to 2024) before the cumulative discovery of two-thirds of the total resources necessarily generates a more rapid decline in output. In this case the possibility of producing more than 23 x 10^9 barrels per year remains until 2035 and the opportunity of producing more than 10 x 10^9 barrels to 2066. By 2080 production is still almost seven x 10^9 barrels and by then almost one-tenth of the world's oil still remains undiscovered.

Figures 4.27, 4.28 and 4.29 present cases with successively higher resource bases of 5000, 8000 and 11,000 x 10^9 barrels, respectively. In these cases the post-2000 decline in production is clearly unnecessary as so little of the total resources of oil are used up by the year 2080, *viz* 65 per cent, 41 per cent and

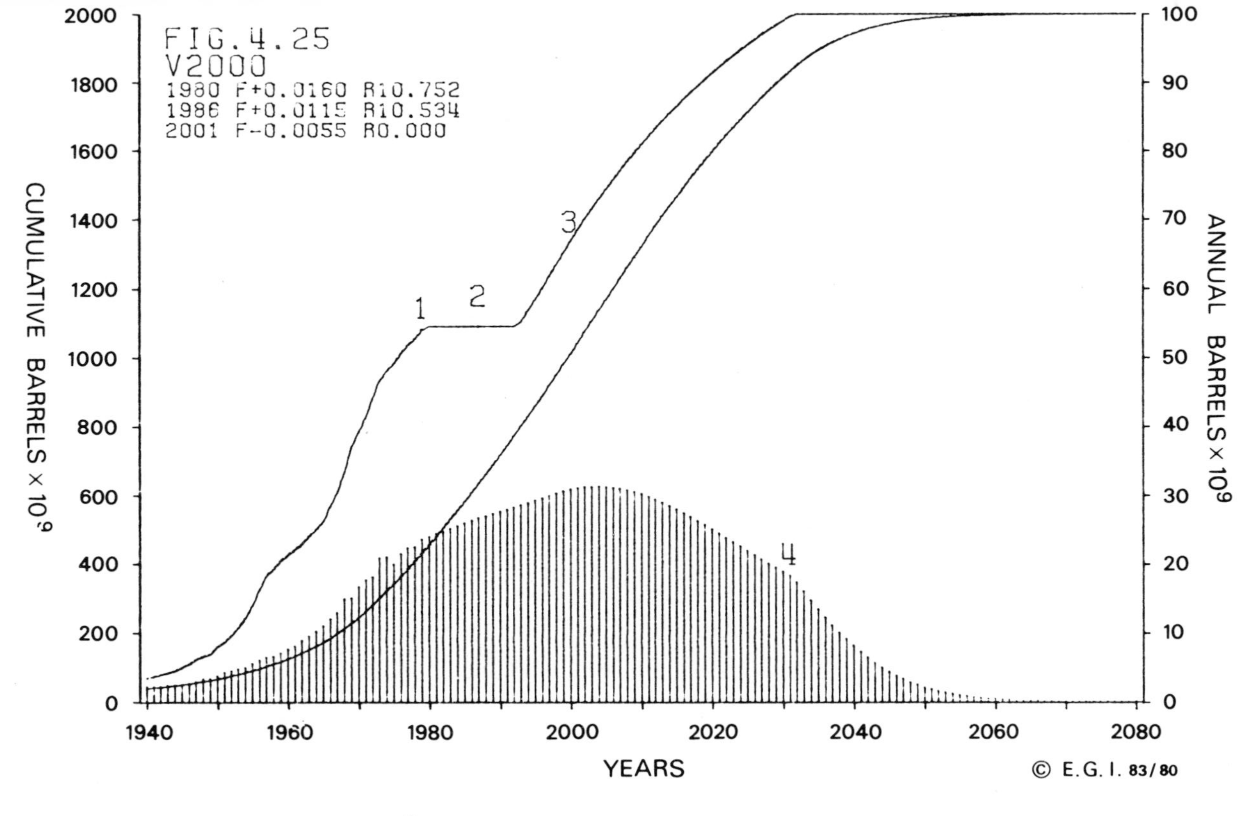

Figure 4.25 *The 2000 x 10^9 Barrels Resource Base, with a 1.6%/1.15%/−0.55% Growth Rate Case*

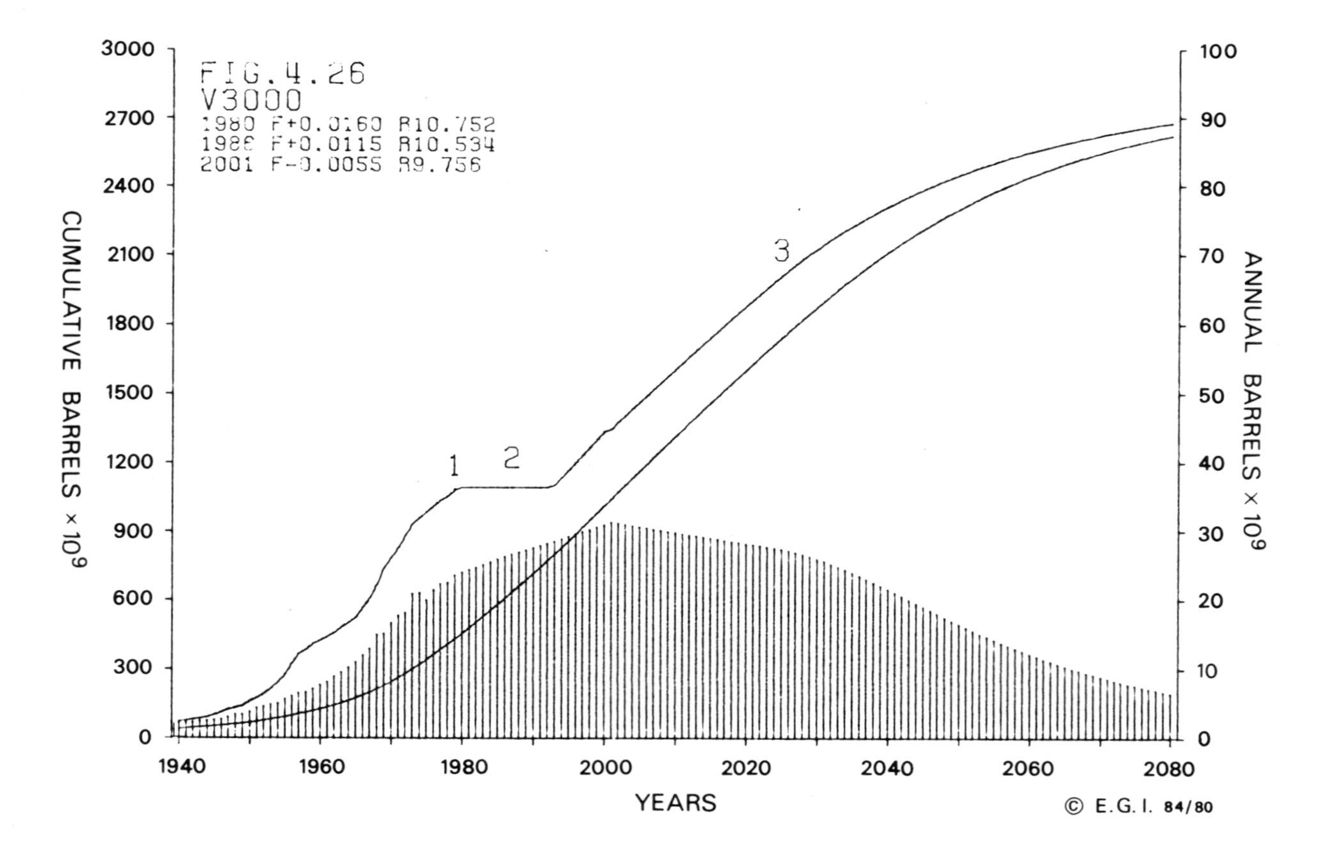
FIG. 4.26
V3000
1980 F+0.0160 R10.752
1986 F+0.0115 R10.534
2001 F-0.0055 R9.756
ANNUAL BARRELS × 10^9
CUMULATIVE BARRELS × 10^9
YEARS
1
2
3
© E.G.I. 84/80

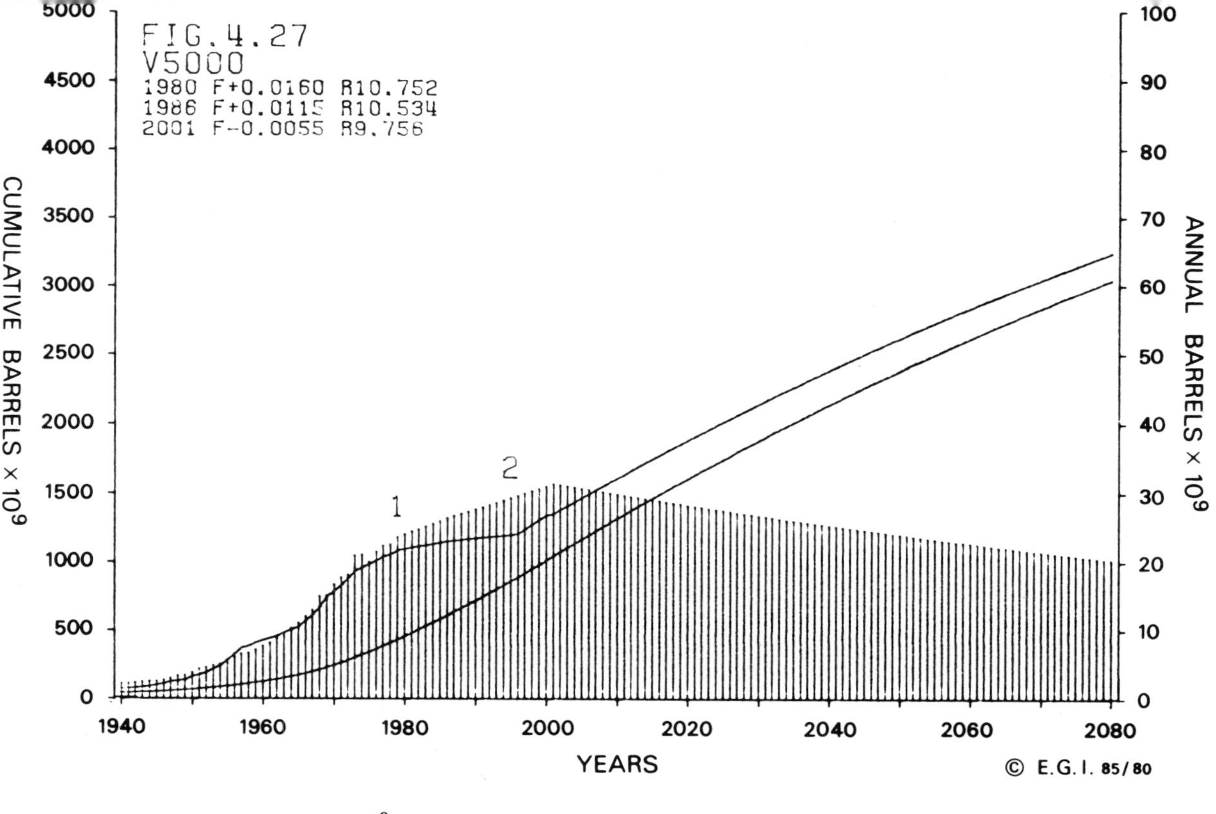

Figure 4.27 *The 5000 x 10^9 Barrels Resource Base, with a 1.6%/1.15%/−0.55% Growth Rate Case*

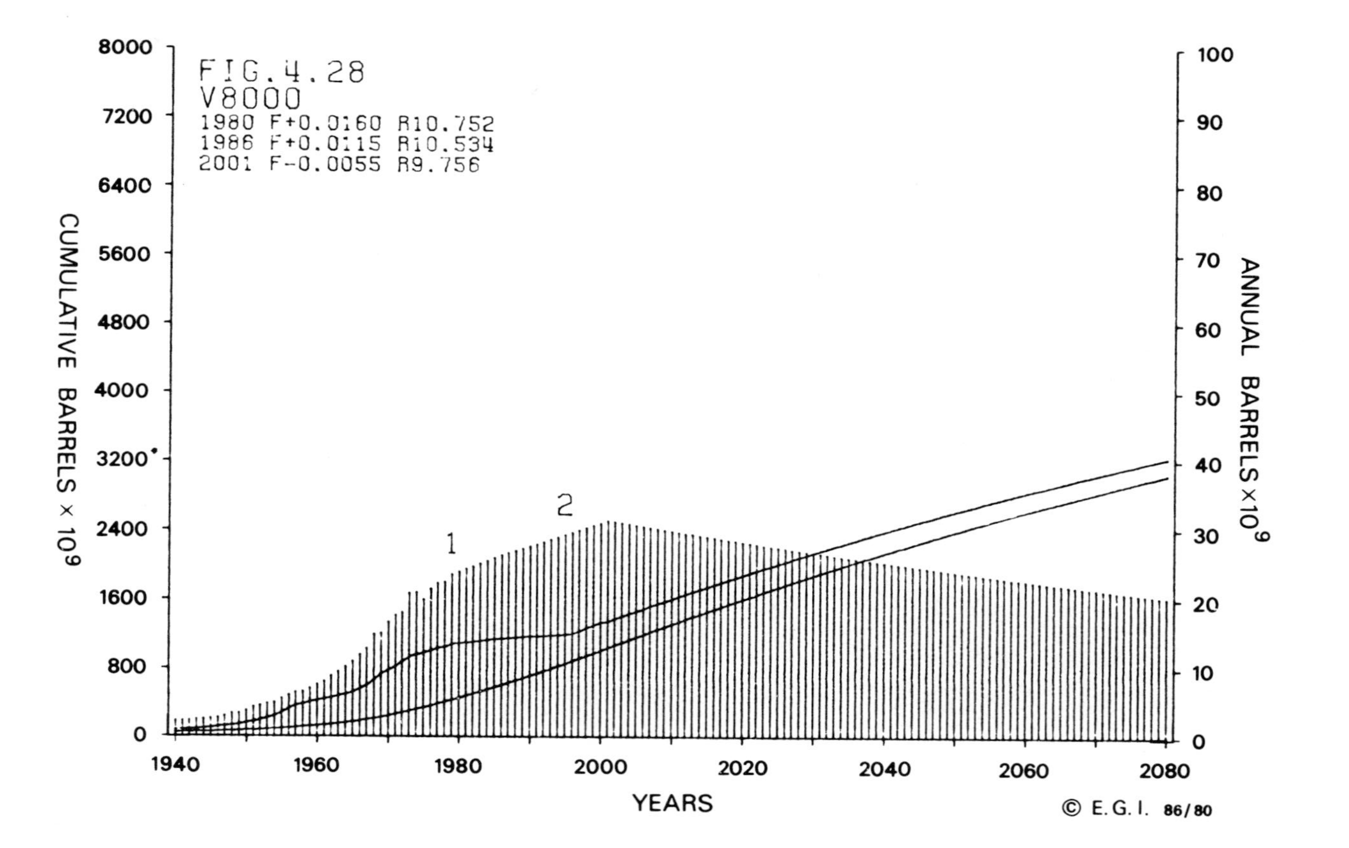
FIG.4.28
V8000
1980 F+0.0160 R10.752
1986 F+0.0115 R10.534
2001 F-0.0055 R9.756
8000
7200
6400
5600
4800
4000
3200
2400
1600
800
0
CUMULATIVE BARRELS × 10⁹
100
90
80
70
60
50
40
30
20
10
0
ANNUAL BARRELS ×10⁹
1
2
1940
1960
1980
2000
2020
2040
2060
2080
YEARS
© E.G.I. 86/80

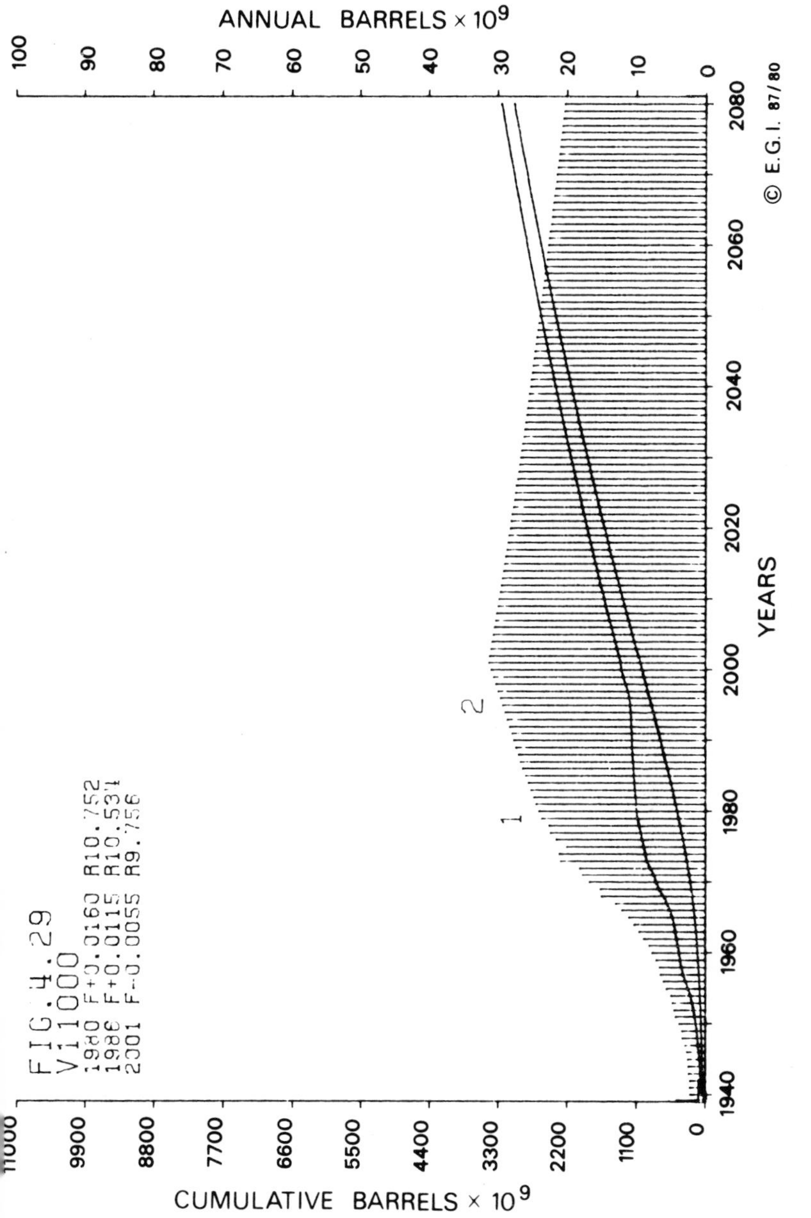

Figure 4.29 *The 11,000 x 10^9 Barrels Resource Base, with a 1.6%/1.15%/−0.55% Growth Rate Case*

29 per cent, respectively. In none of these three cases does the oil industry even begin to approach the time at which a decline in the production of oil is necessitated by the size of the resource base. This, of course, serves to undermine the validity of an exercise which makes a low oil availability restraint so critical for its conclusions on the future of oil. Yet this was the basis on which the Cambridge Energy Group chose to build up its whole approach to the question of the future of oil and it thus excluded some of the options which remain open for the industry's future.

E. Conclusions

There is one essential common implication arising from the three recent industry views of the future of oil that have been presented in this chapter. Their forecasts of a very limited future period of growth in oil are valid only in terms of the most conservative view of world oil resources. Yet, even with the lowest figure of 2000 x 10^9 barrels used in their analyses, the 'tail' of the industry's ability to supply energy to the world is a relatively long one. In all three cases there is certainly enough oil available until the middle of the 21st century to provide for its difficult-to-substitute uses.

On moving to look at the impact of somewhat more extensive resource bases of 3000 and 5000 x 10^9 barrels (resource bases which, as we showed in Chapter 1, involve the enhanced recovery of conventional oil and an assumption that technico-economic developments will make quite modest amounts of so-called non-conventional oil available over the period of the next 20 to 30 years), then the growth curves which BP, Shell and the Cambridge Energy Research Group have built into their models of the future of oil do not generate enough use of the commodity to deplete the world's oil over the period of the next 100 years. And with the larger resource bases – of 8000 and 11,000 x 10^9 barrels – so little of the potential is used that the constrained growth rates, arising from an assumption of physical supply limitations within the next five to 20 years, become entirely inappropriate for a realistic evaluation of the future of oil.

In brief, the companies' presentations represent very special cases in terms of evaluating the contribution of oil to the world's energy economy. They also represent an inexplicable *volte face* from their previously expressed views on the subject (see Chapter 3). The contrast is so marked that the industry's justification for the fundamental change in its outlook is required – and so far that has not been forthcoming. The fundamental change in the industry's views on the long-term future of oil seems to emerge from contrasting short-term, or short-sighted considerations of a kind which, in essence, are not relevant to long-term global oil future studies. In the light of the extreme optimism of the industry followed by extreme pessimism it does not seem inappropriate to set on one side the highly conflicting views of the industry. Instead, a wide-ranging interpretation of the long-term inter-relationships between oil use, supply and resources seems to be required for a good understanding of the alternative futures for oil which are open for possible development over the next 100 years. This is the objective of the following chapter.

Chapter 5

Modelling the Future of Oil with Variable Resources, Reserves and Use Conditions

A. Introduction

In Chapter 2 we defined and described a model which specifies the inter-relationships of oil resources, reserves and use. In Chapter 4 we used the basic framework of this model to test the validity and implications of the recent oil industry views on the future of oil as defined by BP, Shell and the Cambridge Energy Research Group. In essence, the views of these organizations are defined by reference to specified rates of use of oil for a number of periods up to the year 2000 or to 2020. But, as was shown by the tests to which we subjected them, the rates of growth so defined and the periods over which they are to operate are appropriate only for certain, and generally pessimistic, assumptions about the size of the resource base. Neither BP nor Shell, in the studies which we analysed,[1] were specific about their assumptions on the size of the resource base.

Surprisingly none of these views on the future of oil appeared to be related to the possible restraints on production levels which arise from the inability of the industry to find sufficient reserves each year to sustain a specified rate of increase in oil use. This omission, however, could well be a function of the relatively short-term expectations on the industry's growth prospects. Indeed, the highest required annual addition to reserves in the BP, Shell and ERG models is 27, 44 and 35 x 10^9 barrels, respectively. These may be judged to be relatively modest expectations of reserves' expansion for the oil industry in the future, compared with the annual addition of 54 x 10^9 barrels of new and appreciated reserves which were added to the industry's stock of proven reserves each year between 1950

1. *viz* British Petroleum, *Oil Crisis . . . Again?*, London, 1979 and Shell, *Energy Imperatives for the Coming Decades*, Amsterdam, 1979.

and 1970.

In this chapter we present a large set of future of oil outlooks emerging from the inter-relationships of a range of oil resource bases, a range of exponential growth curves for oil use, and the restraints arising from varying maximum possible annual discovery rates of reserves. It is thus, in the many simulations of the future of oil which are presented in this chapter, that full use is made of the model which was described in detail in Chapter 2. The main elements in that model can be briefly re-stated as follows.

The dynamics of the model are provided – principally, though not exclusively – by a defined rate of increase in the use of oil. This operates in any iteration as long as the given size of the resource base makes the defined rate of increase in the use of oil possible. Note, however, that under certain conditions in the period of production build-up, the rate of growth of use may be constrained by a maximum permitted addition to annual reserves. This means that the effective expansion of oil use becomes less than the percentage designated for the iteration. When 66.7 per cent of the defined ultimate resources of oil have been found the rate of additions to reserves begins to fall away (either from the peak figure of annual additions to reserves at that moment or from the maximum annual rate of reserves additions permitted in the iteration) at three per cent per year until all the oil (as defined in the iteration) is discovered; or until 2080, whichever is the earlier. Throughout the model a minimum specified reserves/production (R/P) ratio is required to be maintained in order to safeguard the industry's 10-year future. The calculation of the required R/P ratio is a dynamic one. It reflects the shape of the growth/decline curve in oil use for each year in order to ensure that the next 10 years' oil needs are known to exist as proven reserves. One final point to note, concerning the modelling procedure, is that 2080 is always the termination date, irrespective of the state at that time of the cumulative discovery and production of the total resources in an iteration. Under certain conditions (of resource base size and of rate of increase of use) the oil industry has by no means run its full course by that time. 2080 is, however, so remote in time as to make it pointless to proceed beyond that date in a contemporary presentation on the future of oil.

B. The Iterations of the Model

A total of 40 cases has been run, based on various combinations of input values. Table 5.1 shows the 5 x 5 matrix based on 2000, 3000, 5000, 8000 and 11,000 x 10^9 barrels resource bases and on annual growth rates of 0.65, 1.32, 2.5, 3.3 and 5.0 per cent in the use of oil with a further assumption of no limits on the annual rate of additions to reserves. The output of these 25 cases is presented in Figures 5.1 to 5.25 (and included as Appendix 1 to this chapter).

Table 5.2 presents a second matrix, based on the same ranges of resource bases and growth rates but with the results now incorporating the impact of the additional factor of restricted additions to the annual rate of additions to reserves. This restriction is related to the resource base size. It ranges, as shown in Table 5.3, from a limit of 40 x 10^9 barrels maximum annual additions to reserves with a resource base of 2000 x 10^9 barrels to a limit of 110 x 10^9 barrels of annual reserves additions at the other extreme of an 11,000 x 10^9 barrels resource base. It should be noted, however, that the limits on the annual additions to reserves do not control the development of the future oil in quite a large number of the iterations. This means that, in these cases, the results are identical with those of the same iterations in Table 5.1. The cases are picked out, in Table 5.2, by the rubric, LNA (Limits do not apply). The output for the remaining 15 cases – out of the 25 possible – is presented in Figures 5.26 to 5.40. These diagrams are included in Appendix 2 to this chapter.

C. An Evaluation of the Output

It would be tedious and it is unnecessary to describe the results of each of the 40 iterations. Instead we shall evaluate the results in terms of their main components and implications. From Table 5.1, in which the set of results presented are not constrained by any limit on the annual additions to reserves, we see that the date of the earliest maximum annual production occurs in 2002. This is with an assumption of a resource base of 2000 x 10^9 barrels and an average annual growth rate in oil use of five per cent. From Table 5.4, however, we can see that the rate of growth is constrained below five per cent as early as 1991 – as a result of the control exercised at that date by the discovery of more than two-thirds of the total oil resource base. In the

Table 5.1 *Results of 25 iterations of the model with various combinations of input values and no limit on annual additions to reserves*

Size of Resource Base (x 10^9 bbls) / Percentage Increase in use	0.65%		1.32%		2.5%		3.3%		5.0%	
Key to Tabulation ⇨	Year of Max. Production / Year of Max. Addit. to Reserves*	Volume of Max. Production / Volume of Max. Addit. to Reserves								
2000	2006	27.7	2004	31.5	2004	38.3	2003	42.8	2002	52.3
	2003	29.4	1999	34.9	1995	44.6	1993	51.0	1990	64.5
3000	2026	31.8	2021	39.4	2017	52.9	2014	61.2	2010	77.3
	2024	33.7	2016	43.7	2008	61.5	2004	72.9	1998	95.4
5000	2061	39.9	2047	55.4	2034	80.4	2028	96.4	2020	126.0
	2059	42.3	2042	61.4	2025	93.6	2018	114.8	2008	155.3
8000	post 2080	2080 45.4	2074	78.9	2051	122.4	2041	147.1	2030	205.2
	post 2080	2080 48.4	2069	87.4	2042	142.4	2031	175.1	2018	253.0
11,000	post 2080	2080 45.4	post 2080	2080 88.7	2063	164.6	2051	203.5	2036	275.0
	post 2080	2080 48.4	post 2080	2080 101.0	2054	191.5	2041	242.3	2024	339.0

* In the year following this year the growth rate, as defined, can no longer be maintained because of the fall in additions to reserves. The rate of growth between this date and the date of peak production is thus at a lower rate than indicated at the head of the column.

Table 5.2 *Results of 25 iterations of the model with various combinations of input values and with limits on the maximum annual additions to reserves*

Size of Resource Base (x 10^9 bbls) / Percentage Increase in use	0.65%	1.32%	2.5%	3.3%	5.0%
Key to Tabulation ⇨	Year of Max. Production ¦ Volume of Max. Production Year limit first applies† ¦ Size of limit See** below				
2000	LNA*	LNA	2002 ¦ 36.6 1992 ¦ 40.0 2.29	1999 ¦ 38.0 1990 ¦ 40.0 2.88	1997 ¦ 41.6 1988 ¦ 40.0 3.17
3000	LNA	LNA	2013 ¦ 47.4 2000 ¦ 50.0 2.06	2010 ¦ 48.9 1993 ¦ 50.0 2.34	2006 ¦ 50.8 1989 ¦ 50.0 3.0
5000	LNA	LNA	2030 ¦ 66.8 2013 ¦ 68.0 1.75	2025 ¦ 68.7 2002 ¦ 68.0 1.76	2019 ¦ 70.2 1995 ¦ 68.0 2.54
8000	LNA	LNA	2047 ¦ 92.7 2025 ¦ 92.0 1.44	2041 ¦ 93.7 2012 ¦ 92.0 1.70	2027 ¦ 94.6 1998 ¦ 92.0 2.45
11,000	LNA	LNA	2062 ¦ 111.6 2032 ¦ 110.0 1.28	2047 ¦ 112.1 2017 ¦ 110.0 1.62	2031 ¦ 113.2 2001 ¦ 110.0 2.44

Notes:
* LNA = Limit (of maximum annual addition to reserves) not applicable.
† Impact of limit from this year means that rate of growth in demand after this date is less than that defined for the column.
** Actual average annual growth rate in the period when the limit to annual additions to reserves applies.

Table 5.3 *Resource base size and limits to annual additions to reserves*

Size of Resource Base	Upper Limit to Annual Additions to Reserves ($x 10^9$ barrels)
2000	40
3000	50
5000	68
8000	92
11,000	110

year prior to peak production the growth rate would be less than two per cent.

With the same resource base, however, but with the much reduced growth rate of only 0.65 per cent per annum, the year of peak production is pushed back by only a few years from 2002 to 2006 (see Table 5.1), whilst the size of the peak production is reduced to only just over 50 per cent of that which emerges from a five per cent growth rate (27.7 compared with 52.4 x 10^9 barrels).

At the other extreme, with an assumed 11,000 x 10^9 barrels of resources, there is one iteration with a massive peak year production (in 2036) of 275 x 10^9 barrels – a level of output more than 10 times as high as that of 1979. There are also iterations in which the production peak occurs after the year 2080, as already pointed out. These are in the high resources/ low growth rates iterations, the output for which shows quite modest peak levels of the annual use of oil up to the year 2080 of around 45 x 10^9 barrels. This is less than twice the production level of 1979.

In essence what Table 5.1 clearly indicates is the uncertainty over the future of oil, depending on the view taken as to the size of the resource base and the oil use growth rate. It does, however, also suggest that policy decisions concerned with both the exploration and production of oil, and with the rate at which oil is used, can help to prolong the period for the future expansion of the oil industry. Note that it is only with the smallest resource base figure (2000 x 10^9 barrels) that there is a need for the industry to reach its peak within 30 years (in the years between 2002 and 2006 depending on the rate of growth of oil use). The use of a more realistic figure for the size of the resource base of 3000 x 10^9 barrels – in recognition of the certainty of the future supply of non-conventional oil on a small

Table 5.4 *Part of output from 2000 x 10^9 barrels resource base and a 5 per cent growth case (Figure 5.21) to show reduction of growth rate after year of peak additions to reserves and fall in RP ratio with a declining growth rate*

Year	Production (x 10^9 bbl)	Additions to Reserves (x 10^9 bbl)	Growth Rate** %	R/P Ratio (Years)
1988	36.61	41.54	5.0	12.58
1989	38.44	61.47	5.0	12.58
1990	40.36	64.54 (max.)	5.0	12.58
— —	— — —	— — —	— —	— — *
1991	42.13	62.60	4.93	12.54
1992	43.75	60.73	4.80	12.46
1993	45.22	58.90	4.62	12.36
1994	46.57	57.14	4.39	12.22
1995	47.78	55.42	4.13	12.08
1996	48.86	53.76	3.82	11.91
1997	49.83	52.15	3.49	11.72
1998	50.67	50.58	3.12	11.54
1999	51.40	49.06	2.72	11.32
2000	51.90	47.59	2.35	11.13
2001	52.22	46.16	1.99	10.94
2002	52.35 (max.)	44.78	1.64	10.77
2003	52.32	43.44	1.30	10.61
— —	— — —	— — —	— —	— — †
2004	51.13	25.25	1.01	10.35
2005	48.34	0.0	–0.12	9.95
2006	45.51	0.0	–1.00	9.56
2007	42.70	0.0	–1.89	9.19
2008	39.90	0.0	–2.78	8.84

Notes:
* End of time period T_2 (See Chapter 2, Section D.iii).
† End of time period T_3 (See Chapter 2, Section D.iv).
** Calculated from previous 10 years' growth rate.

scale – generates a range of peak production years from 2010 to 2026, depending on the average annual rate of growth in oil use, in the intervening period. For 5000 x 10^9 barrels of ultimate resources – a figure which allows for as much unconventional as conventional oil to be developed but with the latter still, conservatively, assumed to be on the low side – the growth of the industry could continue, at worst, through to the year 2020 or, with a low average growth rate in demand, to 2061. The resource base figures which are higher than 5000 x 10^9 barrels put back the date of the industry's peak production by a minimum of 50 years into the future and, in some cases, to well over a 100 years (to the period beyond the end date of this simulation study).

In the cases of the higher growth rates in demand and/or the larger resource bases, there is a necessary development of high requirements for the volumes of annual additions to reserves. Historically, the peak year for reserves additions, in terms of declarations made at those times, has been about 75×10^9 barrels. Effective reserves additions, however, exceeded 100×10^9 barrels in several years in the past, if the subsequent appreciation of reserves in specific fields is allocated retroactively to the correct discovery years. The likelihood of future annual additions to reserves rising to a higher level than the figures that have been achieved historically is, in part, a function of the ultimate size of the resource base. This is based on the hypothesis that the more oil there is to be found, the more oil will be found with any given level of exploration effort. In order to incorporate this variable into the model and to enable us to simulate the impact of upper limits to annual discoveries, on the shape of the production and resource depletion curves, the limits (shown in Table 5.3) to annual additions to reserves were used as a further input into the analysis.

The relevant results are, as indicated above, detailed in Table 5.2. It is, however, only in respect of the higher growth rates in demand that the limitations on the rate of annual additions to reserves make any difference to the results (see the cases marked LNA in the Table). In the cases for which the limits are effective, there are two aspects to the results. First, peak production rates are lowered and second, the date of peak production is generally brought forward by a few years. These consequences arise because the upper limit to reserves additions is reached some years sooner than the peak year of unlimited reserves additions. This means that the industry cannot grow as fast or for as long as hitherto expected.

This change has two important effects. First, the defined exponential growth rates in demand cannot be maintained for the whole of the industry's growth period. From about 1990 with the smaller resource bases and later with the bigger resource bases, the rate of growth in oil use in the period of growth remaining after the peak addition to reserves has been achieved, is at a rate which declines from that defined for the iteration. This information is shown in Table 5.2 for each of the 15 iterations. Second, the rate at which the industry declines is significantly slowed down. Examination of the data output from appropriate iterations shows that the dates at which production falls back to the level of 1979 can be delayed by up to 20 years. This gives an extended period in which oil could be expected to maintain a higher share of its contribution to total energy needs of the world economy. Table 5.5 shows part of the output for the two iterations (one without limits to reserves additions and the other with such limits) for 5000 x 10^9 barrels of resources and a 3.3 per cent per annum growth rate in oil use. A comparison of Figure 5.18 and Figure 5.33 shows these differences in graphical form.

D. The Model and the Real World of Oil

Appendices 1 and 2 to this chapter consist of 40 graphs showing various combinations of the principle parameters of the model. The salient points from the 40 cases have been summarized in Tables 5.1 and 5.2 and have been highlighted earlier in the chapter. The graphs are presented in order to enable the reader more easily to compare and consider a wide range of alternative prospects for the future of oil. With this wide range of resource bases and growth rates combining, in various permutations, to produce such a broad range of prospects for oil, the question immediately arises as to how likely it is that the actual and continuing development of the oil industry will resemble any particular one of the cases presented.

Before proceeding to a more detailed assessment of relative possibilities, it must be pointed out that it is extremely unlikely that the future of oil will be faithfully represented by any one of the cases presented here. This is because of a basic assumption in the model: namely, that the world of oil supply and demand will be a fully integrated one, viewed geographically, so that there are equal opportunities for all oil-consuming countries to secure their oil requirements from all producing and potentially producing areas of the world; and reciprocally, for all oil-producing regions to be free to supply the demand for oil from anywhere in the world. This assumption has been made because it represents the most efficient manner for the systems of global oil and energy to function.

In this context we would remind readers of the BP and Shell oil supply/demand studies which were examined in Chapter 4. Both of them were concerned only with the non-communist

Table 5.5 *Selected output from two cases of 5000 x 10^9 barrels resources and 3.3 per cent growth rate, viz A with no limits on annual additions to reserves and B with a limit of 68 x 10^9 barrels on annual reserves additions*

Year	A		B	
	Production	Volume of Reserves Additions	Production	Volume of Reserves Additions
2017	81.0	111.1	66.0	68.0
2018	83.7	114.8*	66.5	68.0
2038	86.3	62.4	59.4	44.4
2039	81.7	19.0†	58.1	43.1
2040	75.9	0.0	56.8	41.8
2052	22.6**	0.0	41.0	28.9
2053	19.8	0.0	39.7	28.1
2054	17.4	0.0	38.5	27.2
2068	2.1	0.0	24.7	17.8
2069	1.8	0.0	23.9	17.2
2070	1.5	0.0	23.2	16.7
2071	1.3	0.0	22.5**	16.2
2072	1.1	0.0	20.7	0.2†
2073	0.9	0.0	18.9	0.0
2079	0.3	0.0	10.6	0.0
2080	0.3††	0.0	9.5††	0.0

Notes:
* Maximum additions to annual reserves (A only).
† Year when last resources are added to reserves (A and B).
** Year when production falls below 1979 level (A and B).
†† Production in final year of simulation (A and B).

world oil supply and demand future and so they implicitly assume that the political division between the communist world and the world of the market economies will continue to be the most important barrier (indeed, the only barrier worth noting). Thus the overall development and use of the world's oil resources will be inhibited.

The logic of oil companies making this assumption for their own internal use is apparent, given the long history and the future expectations of their being prevented from participating in either developing the supply of oil in the communist world or in meeting its demand. The logic of excluding the communist areas from studies which are designed to heighten public awareness of the industry is less easy to understand. There is already a degree of contact over oil between East and West. Russian oil is sold on Western markets whilst contracts for

Western companies' co-operation in the exploration and development of Soviet hydrocarbon resources have been prepared and in several cases actually signed, although their implementation would appear (in mid-1980) to have been delayed as a result of immediate political difficulties. In terms of the time span covered by these long-term oil supply/demand studies, however, political difficulties of the moment are hardly significant and the possibility of Russian oil flowing to Western markets in increasing quantities cannot be excluded. Indeed, should the technical, advisory and service contracts to help develop the oil and gas resources of the communist countries become effective, such flows would be necessary to enable the USSR to earn the hard currency necessary to pay for the contracts.

On the other hand, there are those who believe that the USSR will quite soon be forced to participate in the world oil market as a major importer.[1] This, of course, would have an equally important effect on the non-communist world's oil future. Thus, to ignore the other half of the world in looking at the future of oil is a serious omission.

This specific political influence on the development and use of the world's oil resources has been given here simply as an example of the sort of interference which one can expect to influence the future of world oil. Other examples would be the use by exporting countries of oil as a weapon to enforce their political ideals, for example, by the deliberate curtailment of production, which in itself, contributes to the generation of lower growth rates in demand. Unhappily, it is impossible to model such shocks and traumatic events of the future. And it is because of these unknown factors that we suggest that the future of oil will not closely resemble any one of the pictures we have presented in this study. Nevertheless, the net result of all the uncertainties and vicissitudes which are likely to occur can be expected to fall within the wide range of possibilities presented in this chapter. By comparison, other studies of the future of oil which have been undertaken have been based on a specific, or a very narrow, range of values for each parameter. They are thus so rigid in their structure and their presentation as to be useless in their ability to reflect the inherent uncertainty over the future of oil. In particular they

1. See, for example, reports of the Central Intelligence Agency, *op.cit.* (See Chapter 4, p. 132).

are unable to reflect the changing political situation and other restrictions and opportunities which are likely to be imposed on the development of the oil system.

E. Probable Oil Futures

In order to integrate the 40 'futures of oil' which we have presented in this chapter we would now ask the reader to consider the future of oil as a two-dimensional probability surface, the axes of which consist of growth rates and ultimate resource bases. This is shown conceptually in Figure 5.41. Each of the alternative growth rates and each of the alternative ultimate resource bases is then a cross-section through this probability density surface and our cases are located at the intersections of the appropriate values. If we could estimate this surface we could then speak of the probability of any individual case occurring.

In order to do so let us first decompose this two-dimensional surface into two one-dimensional probability distributions and attempt to assign values, in probability terms. We take first the oil use growth rates and for clarity we shall do the assignments of probabilities cumulatively. At the lowest end of the range of values a rate of 0.65 per cent has been taken. This was chosen because it was the historic average growth rate in oil use in the non-communist world in the years 1973-79. During this period the Western world was more severely affected by the controls instituted on production, and by the order of magnitude increase in oil prices, than was the Eastern block. The latter was cushioned from the full effects of these factors by the energy policy of the USSR, one of the world's major oil producers and exporters. In this low growth rate choice we see policy changes being forced on the USSR, with the result that the communist countries' pattern of change in oil use will come into line with the patterns that are being established in the West. We judge this to be possible and so assign a probability of 0.10 to the growth rate in oil use being 0.65 per cent or less.

The overall world average growth rate in oil use from 1973 to 1978 was 1.32 per cent[1] and our use of this growth rate therefore assumes that the role of oil in the world economy will continue in a way similar to that in the period since 1973. This

1. Although 1979 is now included in the historical section of the model, the research was initially undertaken during that year, prior, that is, to the availability of oil production/use data for 1979. All growth rates were, therefore, calculated on the basis of 1973-78 figures.

would seem to be a rather likely course of development and we thus assign a growth rate in oil use of up to 1.32 per cent per annum, a probability of 0.30.

The growth rate of 2.5 per cent per annum represents a modest increase in the rate of growth of oil use compared with the recent past. It is an increase which could come about by an improvement in the Western world's economic performance but with oil use still being held back by high and increasing prices. We consider this is also a highly likely occurrence and assign an oil use growth rate of up to 2.5 per cent per annum, a probability of 0.82.

The growth rates of 3.3 per cent and five per cent are taken simply as higher rates which would emerge from a return to the *status quo ante* the changed world of oil power in 1973. They are thus not very likely. Consequently, the likelihood of growth rates up to these levels were assigned cumulative probabilities of 0.91 and 0.98, respectively. What remains is an 0.02 probability that there will be a growth rate higher than 5.0 per cent. This is a near unattainable rate in the light of present and expected circumstances.

The reasons for the various estimates of the oil resource base were set out in Chapter 1 so we shall now only briefly recapitulate these. A resource base of more than 11,000 x 10^9 barrels entails the full exploitation of all the world's conventional oil including extensive enhanced recovery procedures together with the exploitation of all unconventional oil. To the likelihood of a resource base of this size or larger we assign a very low probability of only 0.04. A base of 8000 to 11,000 x 10^9 barrels would imply the full recovery of the world's conventional oil plus the more accessible unconventional oil. To a resource base of more than 8000 x 10^9 barrels we assign a cumulative probability of 0.24. A resource base of 5000 to 8000 x 10^9 barrels implies the extensive development of conventional oil but the recovery of only limited quantities of unconventional oil. We assign a cumulative probability of 0.69 to the oil resource base being more than 5000 x 10^9 barrels. The probability of there being at least 3000 x 10^9 barrels of recoverable conventional and unconventional oil is extremely high and a cumulative probability of 0.95 is assigned to this level of resources. A resource base of at least 2000 x 10^9 barrels is so generally accepted as a minimum that we assign to it a cumulative 0.99 probability. This leaves 0.01 as the probability of a world oil resource base of 2000 x 10^9 barrels or less

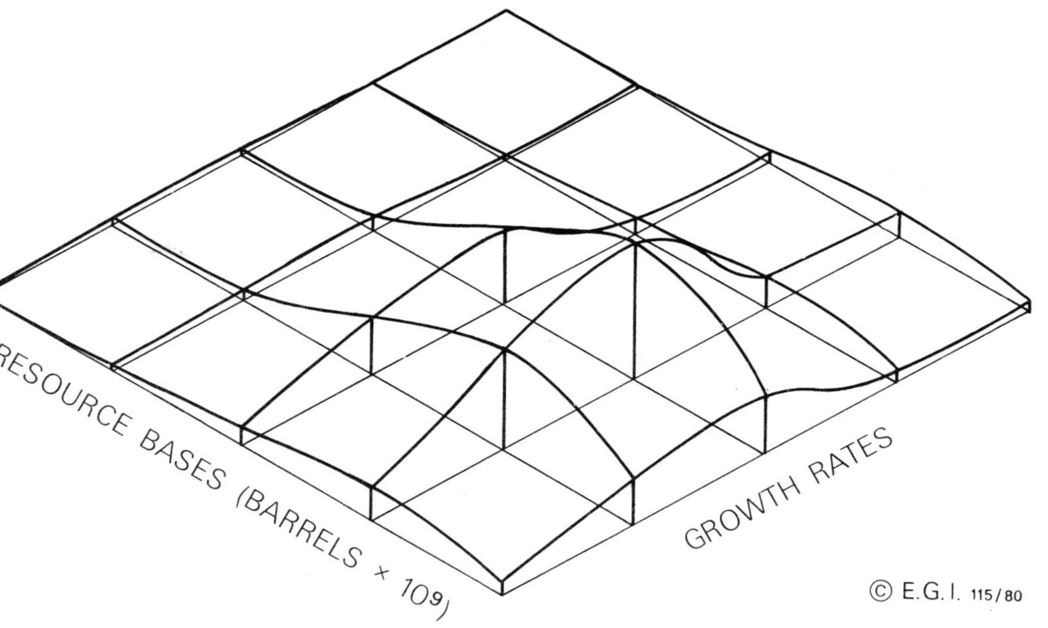

Figure 5.41 *Perspective Drawing of a two-dimensional Probability Surface*

In this conceptual diagram the cross-sections are intended to represent the two main variables used in the model, viz. growth rates in oil use and the size of the oil resource bases. The height of the surface at each intersection gives the combined probability for a specific case.

Note: Figures 5.1 to 5.40 are in Appendices 1 and 2 to this chapter.

Before proceeding further in the analysis we must clearly state that these probabilities are our subjective assessments. They do not emerge from the study in any mathematical way and are based solely on our judgements. They follow from our expectations based on our knowledge and understanding of the industry. The probabilities of each interval are given in Table 5.6 and the absolute and the cumulative probability curves on which they are based are shown in Figures 5.42 A and B and 5.43 A and B.

Table 5.6 *Subjectively assigned probabilities by interval – growth rates and resource bases*

Growth Rates		Resource Base	
Interval	*Probability*	*Interval*	*Probability*
≤ 0.65	0.10 (0.10)*	> $11,000 \times 10^9$	0.04 (0.04)
> 0.65–1.32	0.30 (0.40)	> 8000×10^9 – $11,000 \times 10^9$	0.20 (0.24)
> 1.32–2.50	0.42 (0.82)	> 5000×10^9 – 8000×10^9	0.45 (0.69)
> 2.50–3.30	0.09 (0.91)	> 3000×10^9 – 5000×10^9	0.25 (0.94)
> 3.30–5.00	0.07 (0.98)	> 2000×10^9 – 3000×10^9	0.05 (0.99)
> 5.00	0.02 (1.00)	≤ 2000×10^9	0.01 (1.00)

*The numbers in brackets are the cumulative probabilities – as given in the text.

It is now possible to calculate the probability that the future of oil will lie in any one of the intervals of the joint probability distribution or in any combination of intervals. This is based on the data from Table 5.6. The probabilities of specific sectors of the two-dimensional probability distribution are then indicated in Table 5.7. Each entry in this Table indicates the probability of a specific combination of growth rate and resource base. For example, the combination which emerges out of a growth rate of more than 1.32 per cent and less than or equal to 2.5 per cent and a resource base of between 5000×10^9 and 8000×10^9 barrels has a calculated probability of 0.1890. This is the peak probability for any cell of this distribution. All cell entries are presented as non-cumulative probabilities as this allows the calculation, by the reader, of the probability of each of the cases which have been presented in this chapter.

Cumulative probabilities have been calculated for four sectors of the distribution in Table 5.7. These sectors are identified by the numbers 1, 2, 3 and 4 joined by the dashed lines. These cumulative probabilities for the top right-hand cell in each sector are listed on the line below the Table (1 = 0.0960; 2 = 0.5658 etc).

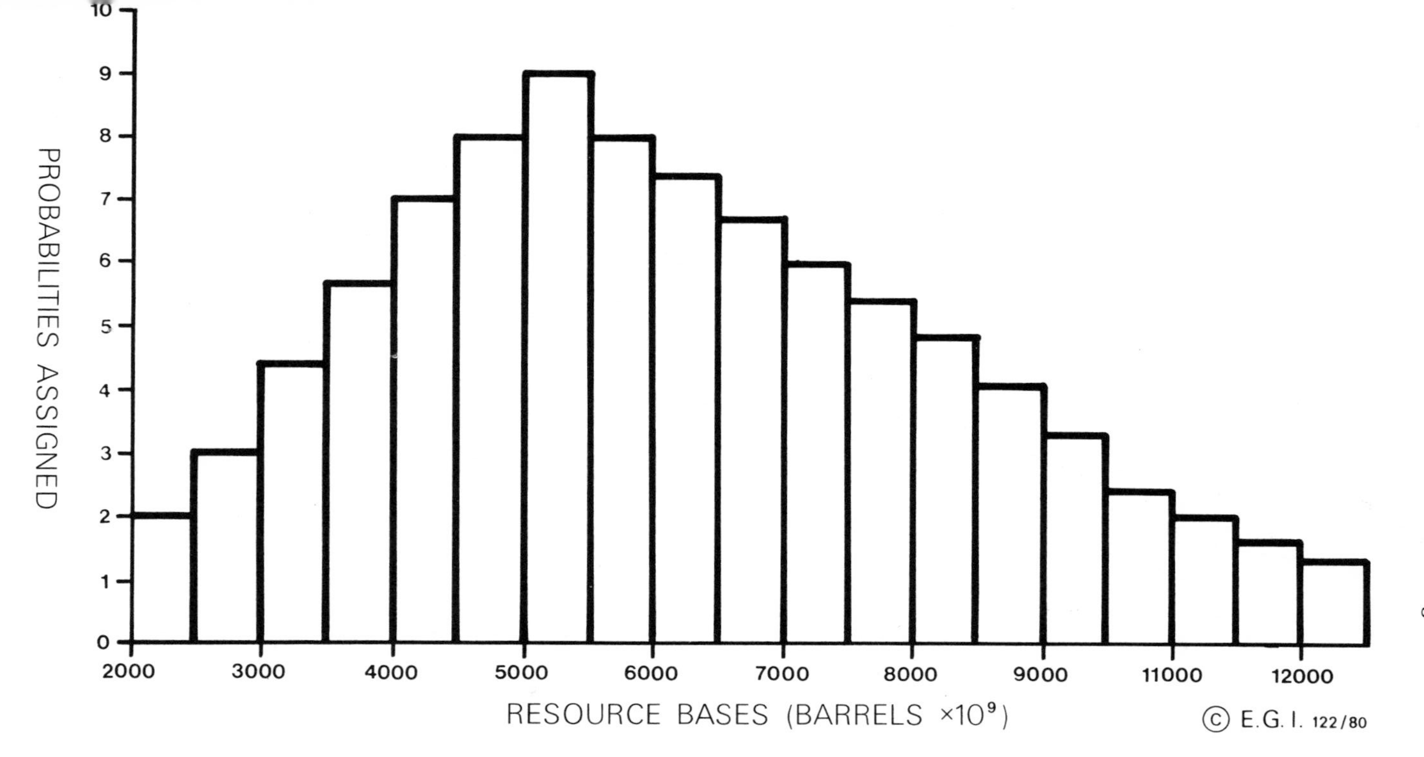

Figure 5.42a *Absolute Probabilities Assigned to Various Resource Bases*
The height of each line shows our assessment for the probability of each base.

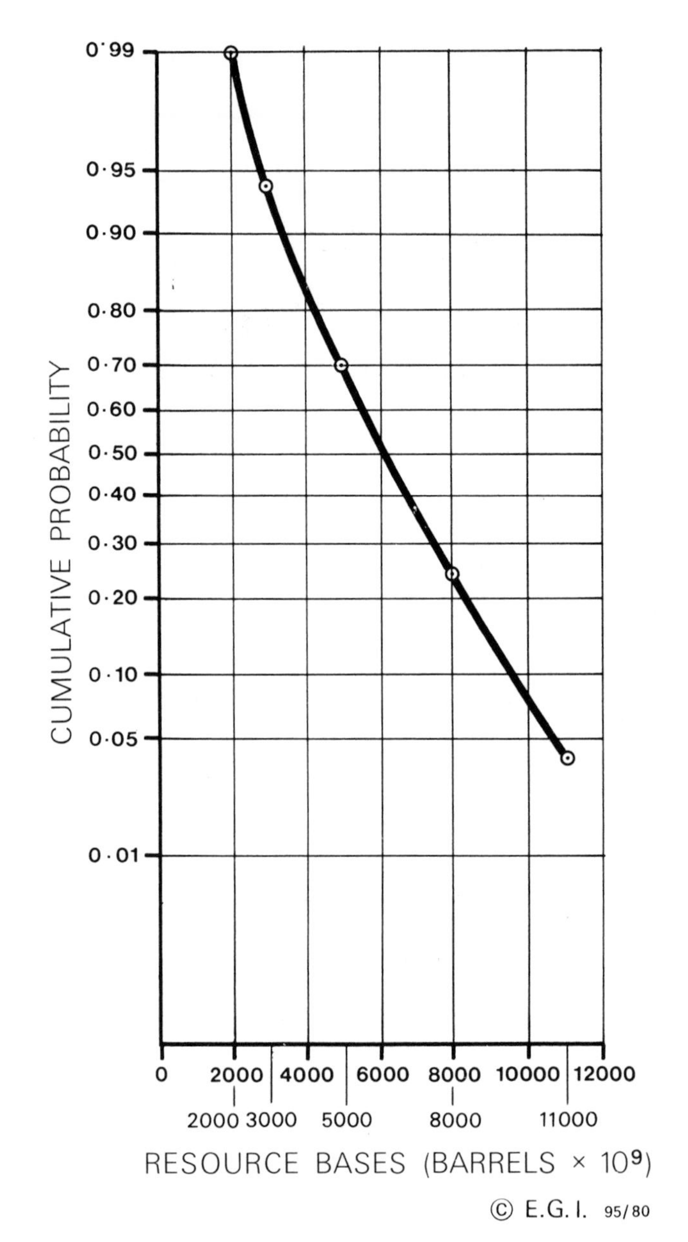

Figure 5.42b *The Cumulative Probabilities of the Resource Bases*
Each point on this curve represents the probability (Y-axis) of the resource base being as large as or larger than the appropriate value (X-axis).

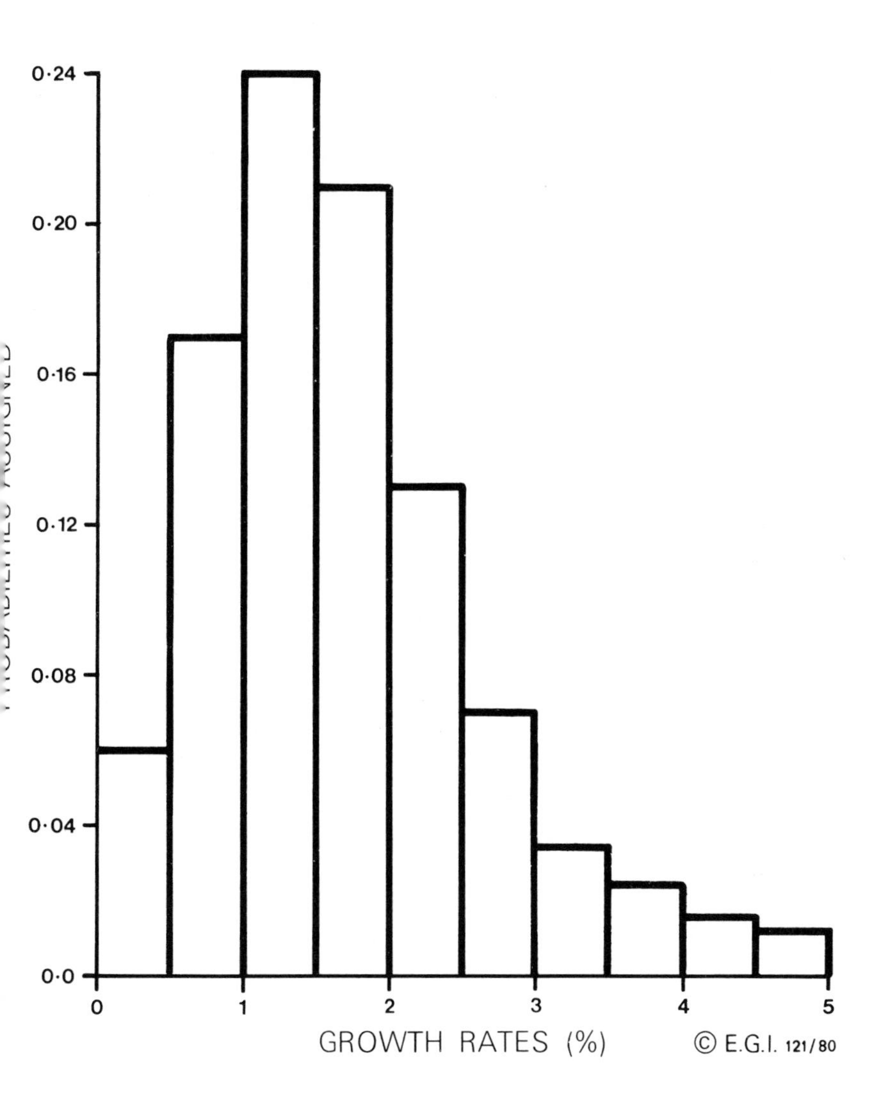

Figure 5.43a *The Absolute Probabilities Assigned to Various Growth Rates*

The height of each line shows our assessment for the probability of each growth rate.

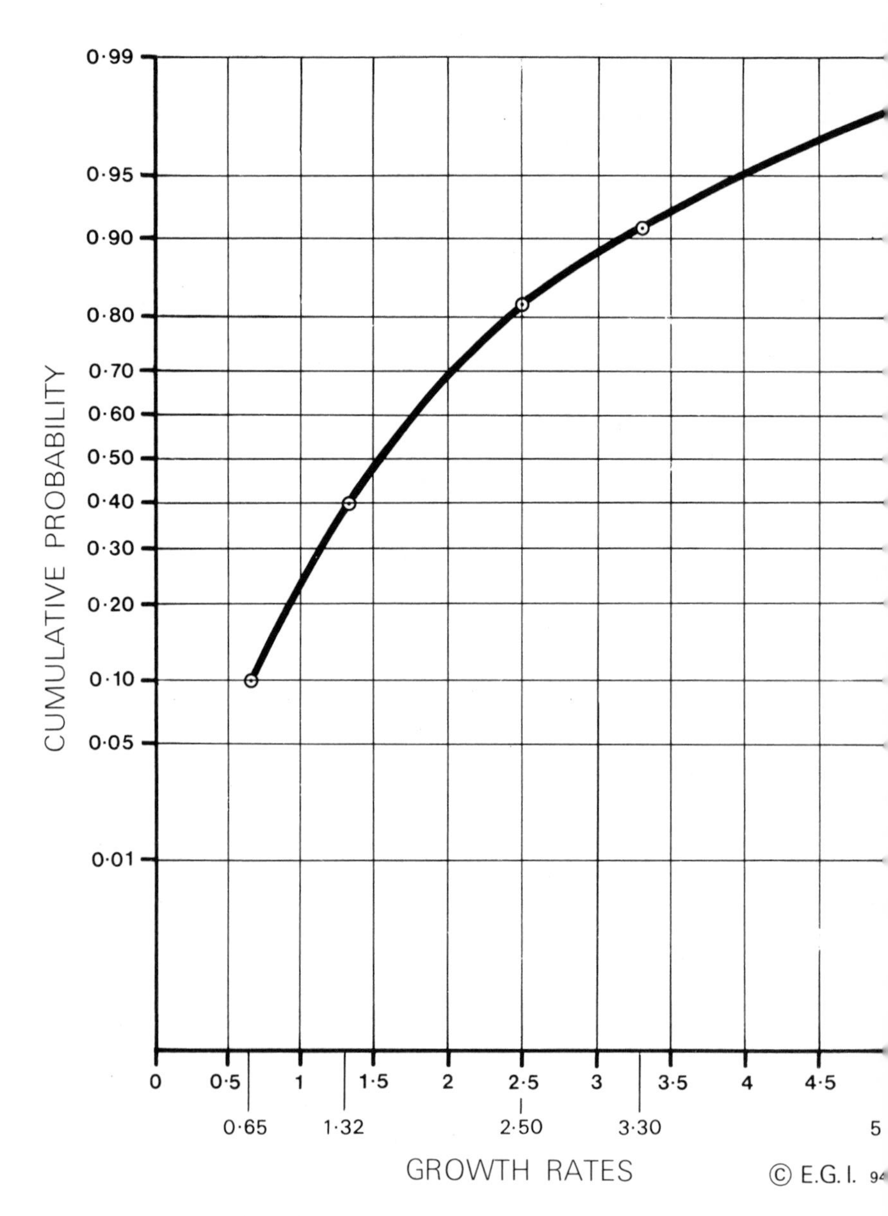

Figure 5.43b *The Cumulative Probabilities of the Growth Rates*
Each point on this curve represents the probability (Y-axis) of the growth rate being as low as or lower than the appropriate value (X-axis).

Table 5.7 *Joint probabilities of the intervals**

Size of Resource Base (x 10^9 barrels)	Growth rates (%)					
	≤ 0.65	> 0.65 – 1.32	> 1.32 – 2.50	> 2.50 – 3.30	> 3.30 – 5.00	> 5,00
≤ 2000	0.0010	0.0030	0.0042	0.0009	0.0007	0.0002
> 2000–3000	0.0050	0.0150	0.0210	0.0045	0.0035	0.0010
> 3000–5000	0.0250	0.0750	0.1050	0.0225	0.0175	0.0050
> 5000–8000	0.0450	0.1350	0.1890	0.0405	0.0315	0.0090
> 8000–11,000	0.0200	0.0600	0.0840	0.0180	0.0140	0.0040
>11,000	0.0040	0.0120	0.0168	0.0036	0.0028	0.0008

1 = 0.0960, 2 = 0.5658, 3 = 0.8554, 4 = 0.9702

* For computational purposes the probabilities assigned to the intervals (Table 5.6) are expressed to two decimal places and in this Table to four decimal places. This apparent high degree of accuracy is, of course, spurious and in working further with this Table the true lower level of accuracy must be kept firmly in mind.

We now apply a value judgement to Table 5.7. The upper right-hand corner of this Table we define as the 'worst' future for oil and the lower left-hand corner we define as the 'best'. In doing this we thus make explicit our judgement that a large oil resource base is 'better' than a small resource base and that a low growth rate in oil use is 'better' than a high growth rate. These views may be disputed – perhaps more so with regard to the demand growth rate than with the resource base size – but it is our considered judgement that the long-term future of the world's economy and of the oil component of that economy are best served by a combination of low demand rate and high resource availability.

Our view that the upper right-hand corner of this Table constitutes the 'worst' future for oil can be extended to each of the sectors of the distribution. The division of the distribution is identified by the lines numbered 1, 2, 3 and 4. The upper right-hand corner of each sector is the 'worst' case[1] of that sector. Each of these locations, moreover, occupies a position in

1. Worst case analysis is a commonly used procedure in the field of Operations Research. When combinations of possible results are very high, only the worst possible result is thoroughly analysed since all other results must be better.

the Table which represents one of the cases shown in Table 5.1. Thus, the top right-hand corner of sector 1 corresponds to the case shown as Figure 5.22; the top right-hand corner of sector 2 corresponds to the case shown as Figure 5.18; the corner of 3 to Figure 5.14; and the corner of 4 corresponds to Figure 5.10.

We shall now continue this explanatory discussion using the sector identified as number 2, and the case corresponding to that in the top right-hand corner of that sector (shown as Figure 5.18) as an example. This case has a cumulative probability of 0.5658 and is the sum of the probabilities of all the cells included in sectors 1 and 2 (*viz* 0.045 + 0.135 + 0.189 + 0.020 + 0.060 + 0.084 + 0.004 + 0.012 + 0.0168 = 0.5658); that is, to the sum of the probabilities of the growth rate being less than or equal to 2.5 per cent and of a resource base of, at least, 5000 x 10^9 barrels.

Since the case represented by Figure 5.18 (and corresponding to the location at the top right-hand corner of sector 2 of Table 5.7) is the 'worst' case of sector 2, then all the other cases which occur in sectors 2 and 1 are 'better'. There is thus a probability of 0.5658 that the future of oil will be better than that displayed in Figure 5.18.

The probability 0.5658 is, however, deceptively accurate. Moreover, it is not necessary to use it, as it is possible, using Figures 5.42 B and 5.43 B, to calculate growth rates of demand and sizes of resource bases to give any specific levels of probability desired. We propose to do this for the 0.1, the 0.5 and the 0.9 levels of probability.[1]

F. The 10 per cent, 50 per cent and 90 per cent Probabilities of the Future of Oil

The 0.1 probability case – that is, the 10 per cent likely future of oil – lies very close to the point identified as 1 in Table 5.7. The actual individual probabilities involved are 0.25 for the resource base and 0.41 for the growth rate yielding a joint probability of 0.1025. The resource base of the 0.1 probability case can then be read from Figure 5.42 B as 7900 x 10^9 barrels and the growth rate from Figure 5.43 B as 1.34 per cent. The graph of this case is presented as Figure 5.44 and shows production peaking in 2072 at a level of 78.4 x 10^9 barrels of

1. Note the reduction in accuracy when referring to the probabilities not as 0.900 but rather as 0.9.

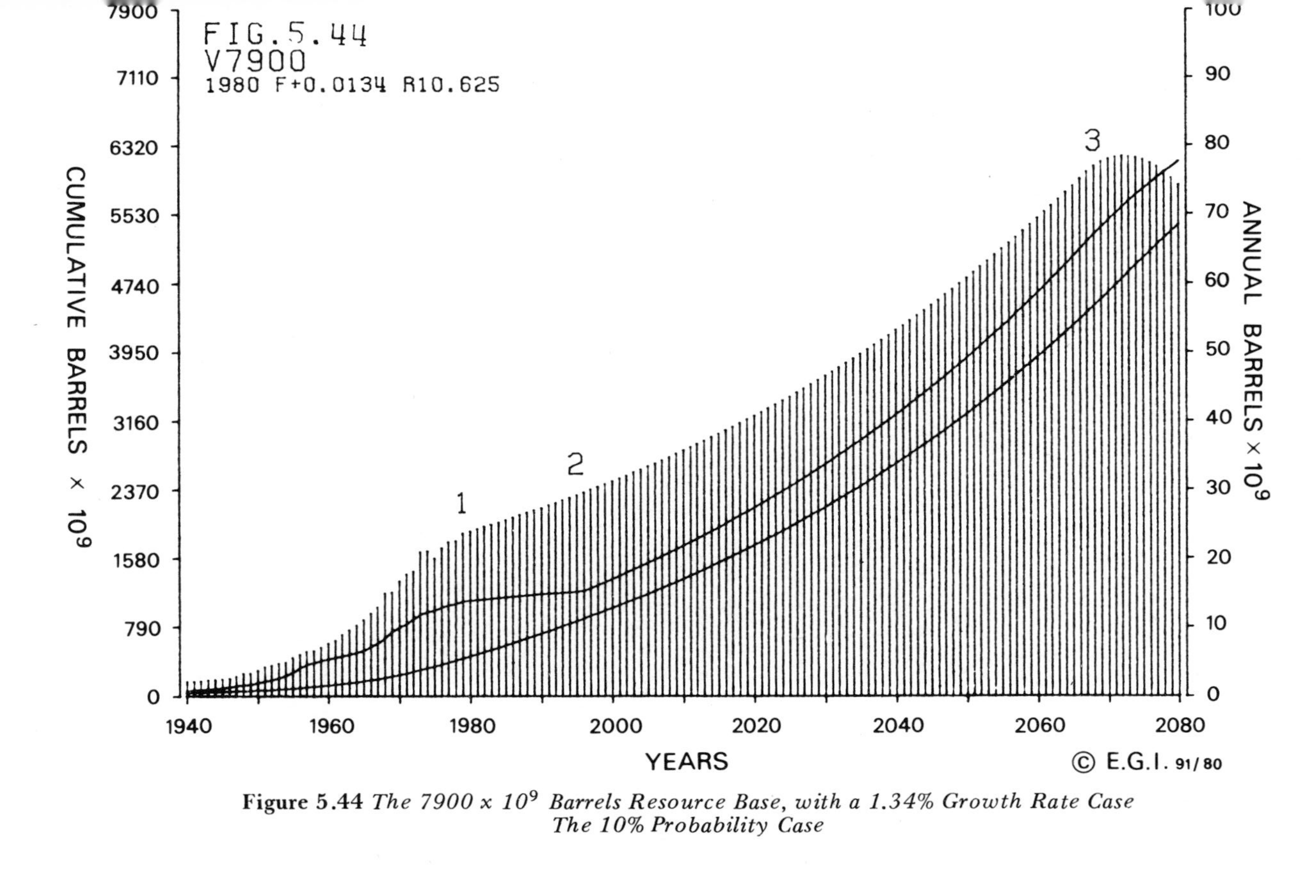

Figure 5.44 *The 7900 x 10^9 Barrels Resource Base, with a 1.34% Growth Rate Case*
The 10% Probability Case

oil. It also shows that 80 per cent of the ultimate resource base has been discovered by 2080. Cumulative production up to the end of the period of the simulation is less than 68 per cent of the resource base. The maximum required addition to reserves, 87 x 10^9 barrels, comes in 2067 after which it is not possible for the resource base to sustain the 1.34 per cent exponential growth curve. Between then and the peak year of production (five years later) the average annual rate of growth in use falls to 1.11 per cent. Given the size of the resource base (at 7900 x 10^9 barrels), an annual limitation to reserves discovered is not deemed to be necessary as the peak required addition to reserves is less than the maximum achievable – 90 x 10^9 barrels as shown in Figure 2.1.

The outcome provides a very relaxing picture of the future of oil. There is continued moderate growth of the industry for nearly 100 years and, even at the end of the study period in 2080, a production level over three times that achieved today can be sustained. We should, however, remember that this is a future of oil at only the 0.1 probability level. There is a 90 per cent chance that the industry's future will not be as good as the future just described.

The 0.5 probability case lies to the left and below the point identified as 2 in Table 5.7. The individual probabilities associated with this point are 0.65 for resource base size which, from Figure 5.42 B, is equal to a base of at least 5200 x 10^9 barrels, and, for the oil use curve, a probability of 0.78 which yields, from Figure 5.43 B, a growth rate of 2.30 per cent. The actual joint probability is 0.5070. Figure 5.45 shows the output from this case. Production peaks in 2036 and 2037 at 77.7 x 10^9 barrels. The maximum annual addition to reserves comes in 2028 at 89.9 x 10^9 barrels and thereafter the rate of growth in use has to fall away. Over the remaining eight years to peak production this rate of growth falls to an average of 1.7 per cent per annum. Production continues for over 30 years beyond the peak, at annual rates in excess of current levels, and it is not until 2069 that production falls below 23.6 x 10^9 barrels. This is eight years later than the last additions to recoverable reserves which occur in 2061.

With a resource base of 5200 x 10^9 barrels it is necessary, following the logic of the earlier portion of this chapter (see above pp. 177 to 178), to define 91.8 x 10^9 barrels as a larger annual addition to reserves than the industry would be able to generate. A limit to the annual increment of reserves of 69.8 x

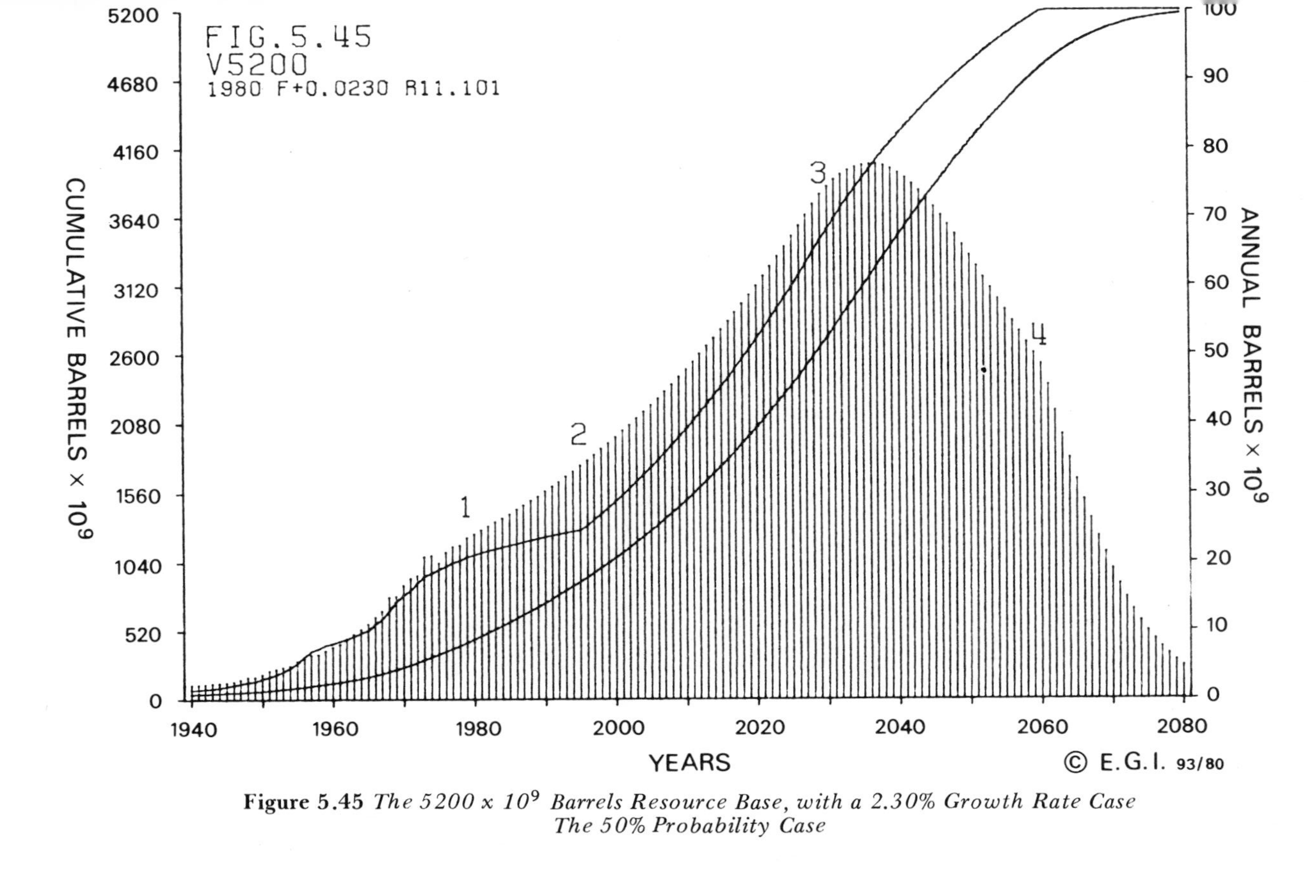

Figure 5.45 *The 5200 x 10⁹ Barrels Resource Base, with a 2.30% Growth Rate Case The 50% Probability Case*

10^9 barrels may be read from Figure 2.1 as the appropriate value for a resource base of 5200 x 10^9 barrels.[1] This 0.5 probability case was now re-run employing this limit. Figure 5.46 displays the resultant graph. The peak production of 68 x 10^9 barrels now occurs in 2033 and in the years from 2017 until 2030 the annual addition to reserves is at the allowed maximum of the 69.8 x 10^9 barrels. This limits the growth of production and shaves almost 10 x 10^9 barrels off the peak, compared with the unlimited case, and represents a 13 per cent reduction in the peak production figure. On the other hand, the introduction of the limit to reserves additions has the effect of filling up the tail of the production curve by creating a less steep rate of decline of the industry. Indeed, in this case, it is not until 2077 that production falls below that of 1979 to give almost a century with an oil industry larger than that of today. At the end of the simulation period (2080) slightly more than 95 per cent of the world's oil resources have been used so that the final demise of the industry would not be long delayed after 2080.

The 0.9 probability level lies very close to the point identified by the number 3 in Table 5.7. The individual probabilities for this 90 per cent probable future of oil are 0.95 for the size of the resource base and 0.94 for the growth rate of demand. Reading from Figure 5.42 B we see that this gives a resource base of 2900 x 10^9 barrels and from Figure 5.43 B that the growth rate required is 3.75 per cent. The actual joint probability is 0.8930. This 0.9 probability case is shown in Figure 5.47. In this, the peak production of 63.0 x 10^9 barrels comes in the year 2012 whilst 2001 is the year of maximum addition to reserves at a level of 75.8 x 10^9 barrels, after which the rate of increase in use has to be less than the 3.75 per cent per annum specified in the simulation. Between then and the year of peak production the annual rate of increase in oil use is 2.8 per cent. While this case represents a considerable deterioration from the favourable picture of the future of oil presented in Figure 5.45, showing the 50 per cent probable case (unconstrained by limits to reserves addition), it still indicates continued growth for the industry for a period of another 32 years. Indeed, it is not until 2030, 50 years in the future, that annual production falls back to a level below that of today. It is important to remember that this 0.9 probability case means

1. See above, Chapter 2, p. 51.

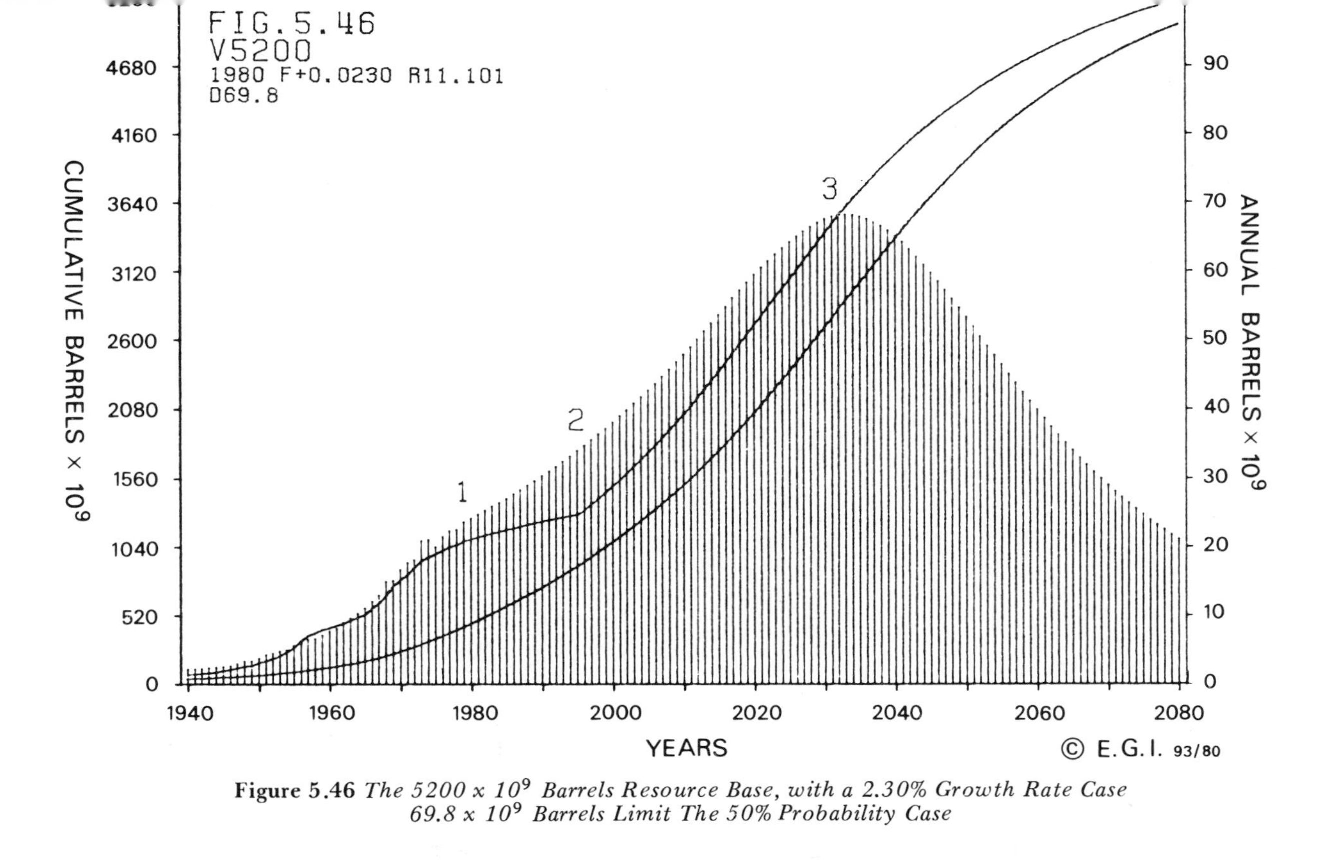

Figure 5.46 *The 5200 x 10^9 Barrels Resource Base, with a 2.30% Growth Rate Case 69.8 x 10^9 Barrels Limit The 50% Probability Case*

that there is only a 10 per cent chance that the future of oil will actually be worse than this.

However, a required addition to reserves of up to 76.0 x 10^9 barrels implies an unreasonably high reserves addition rate for a case in which the resource base is only 2900 x 10^9 barrels. A 3000 x 10^9 barrels resource base involves a limit to annual additions to reserves of 50 x 10^9 barrels per year.[1] This case, therefore, has been re-run with 50 x 10^9 barrels as the maximum allowed annual additions to discovered reserves and is presented in Figure 5.48. Again, the discovery rate limitation shaves the production peak by reducing it to 47.8 x 10^9 barrels, a reduction of some 24 per cent. The limitation begins to apply in 1990 and thereafter it restricts both reserves' growth and production until production peaks in 2011 – just one year earlier than in the unconstrained model. The oil that is 'saved' by the lower peak does, however, dramatically change the production decline curve. This is evident from a comparison of Figures 5.47 and 5.48. Indeed, with a limitation on additions to annual reserves, production remains above the level of 1979 until 2039 – a further nine years compared with the unrestrained case.

G. Conclusion

Consideration of these five probability determined cases and, indeed, of all 45 cases presented in this chapter, indicate that a short-term crisis in oil supply/demand relationships is extremely unlikely. With the exception of the impact of a combination of the smallest resource base with the highest growth rates – events which are both individually, and, even more so, jointly, unlikely – the development of the oil industry need not be constrained until the 21st century. The 50 per cent probability case shows, furthermore, that the medium-term future of oil can also be secure and there is even a 10 per cent chance that the oil supply problem is one which only becomes significant in the long-term future.

This does not mean that there will not be a future oil crisis based on factors other than resources, reserves and the cumulative use of oil but it does mean that efforts to avoid a crisis are worthwhile because there is nothing inevitable about its development. The conclusions of our analysis appear to demonstrate that there is a breathing space available in which

1. See above, pp. 177 to 178 and Chapter 1, pp. 42 to 45.

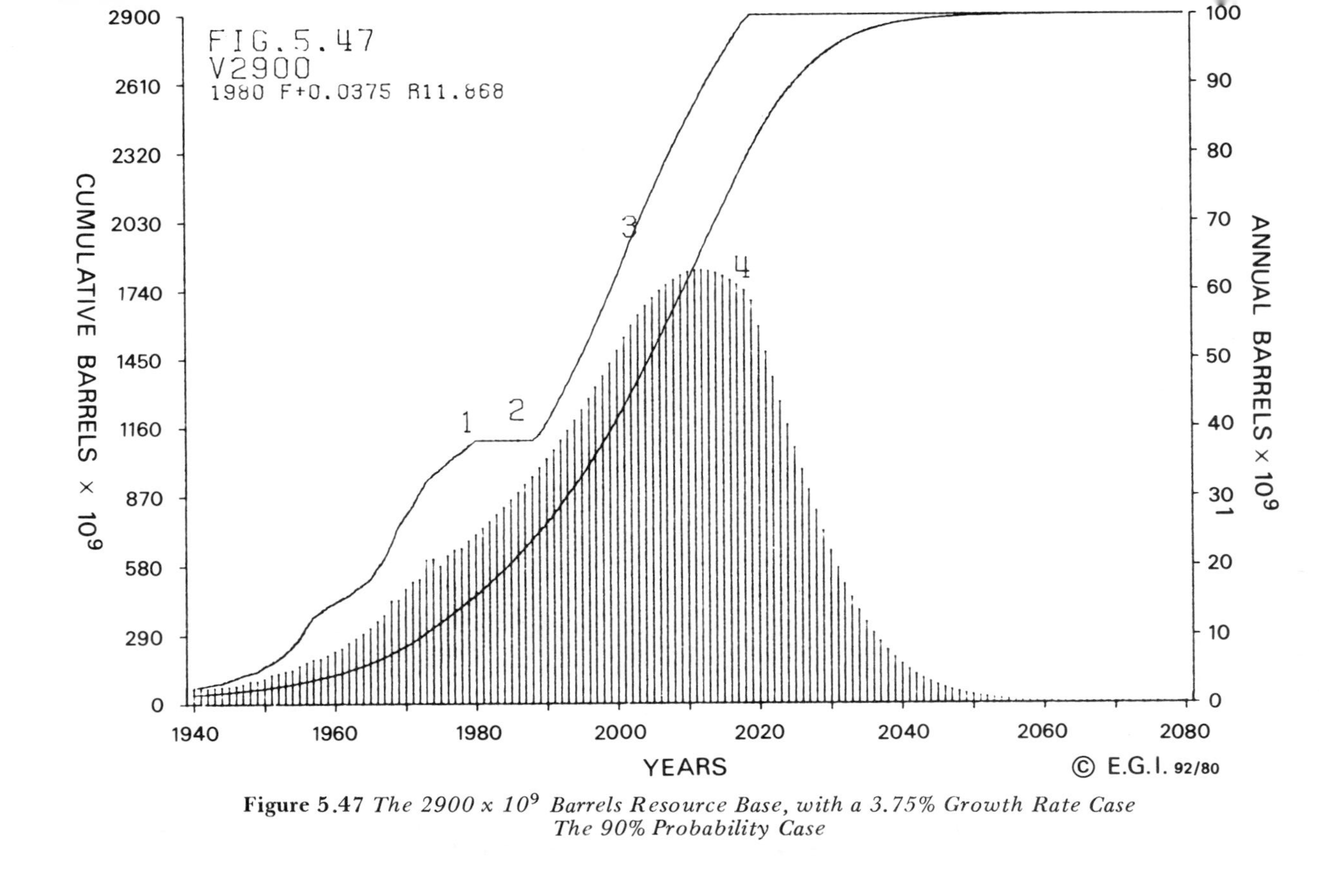

Figure 5.47 *The 2900 x 10^9 Barrels Resource Base, with a 3.75% Growth Rate Case The 90% Probability Case*

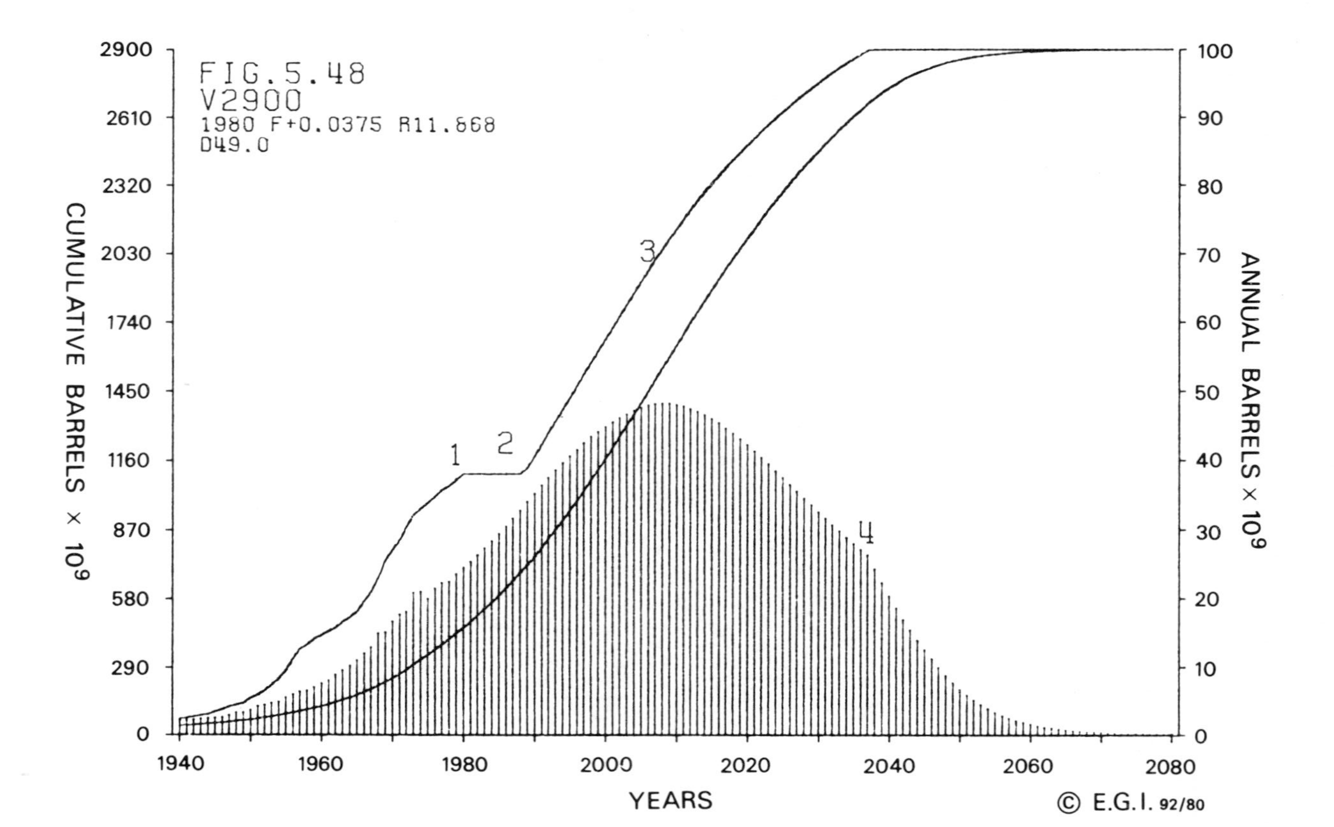

Figure 5.48 *The 2900 x 10^9 Barrels Resource Base, with a 3.75% Growth Rate Case 49 x 10^9 Barrels Limit The 90% Probability Case*

to develop those alternative energies which are certain – some day – to be necessary components in the overall world energy system. Meanwhile, we would emphasize that this conclusion does not mean that in the short- and/or medium-term there will not be oil supply problems. As we have indicated elsewhere such difficulties will arise, but they will be problems arising from political and economic considerations. They will not be the result of the physical lack of enough oil to keep more and more of the world's wheels turning for at least another 30 years. The importance of making sure that energy/oil policies reflect these contrasting considerations on the future of oil will be discussed in the final chapter of this study.

Appendix 5.1

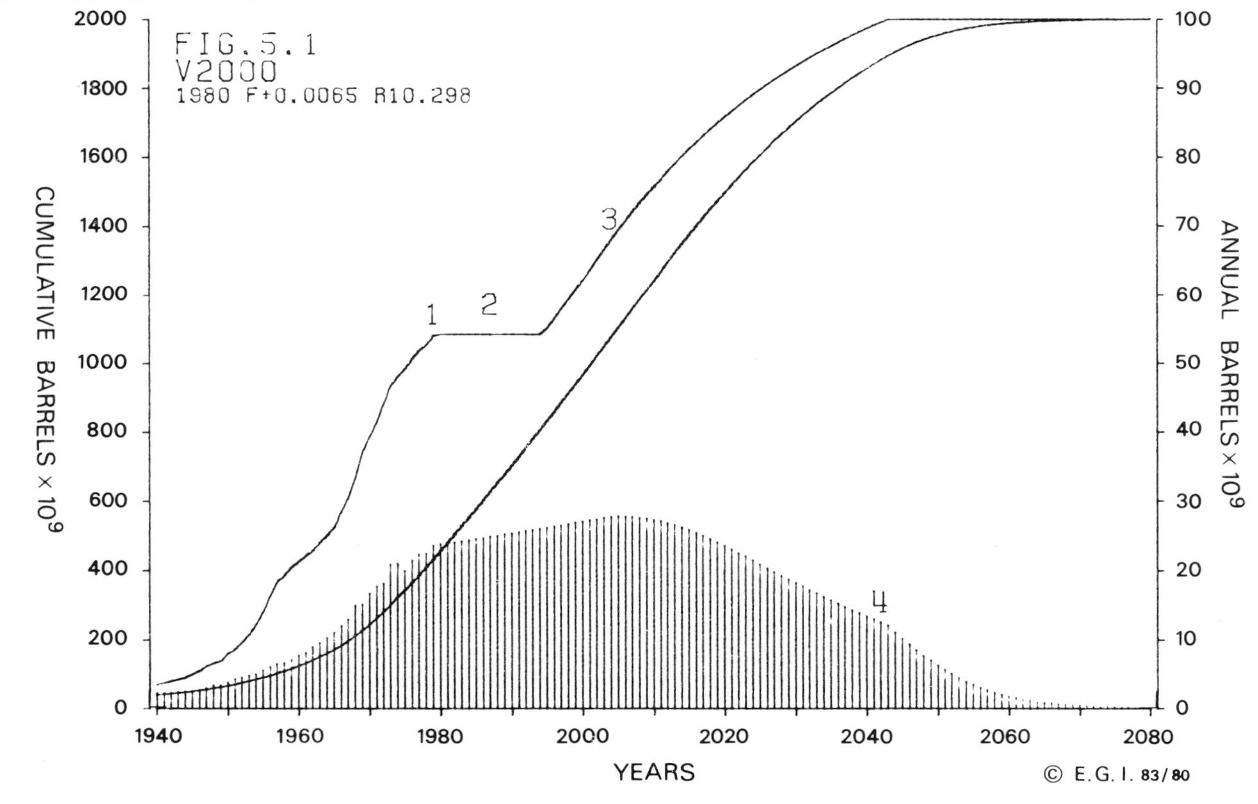

Figure 5.1 *The 2000 x 10^9 Barrels Resource Base, with a 0.65% Growth Rate Case*

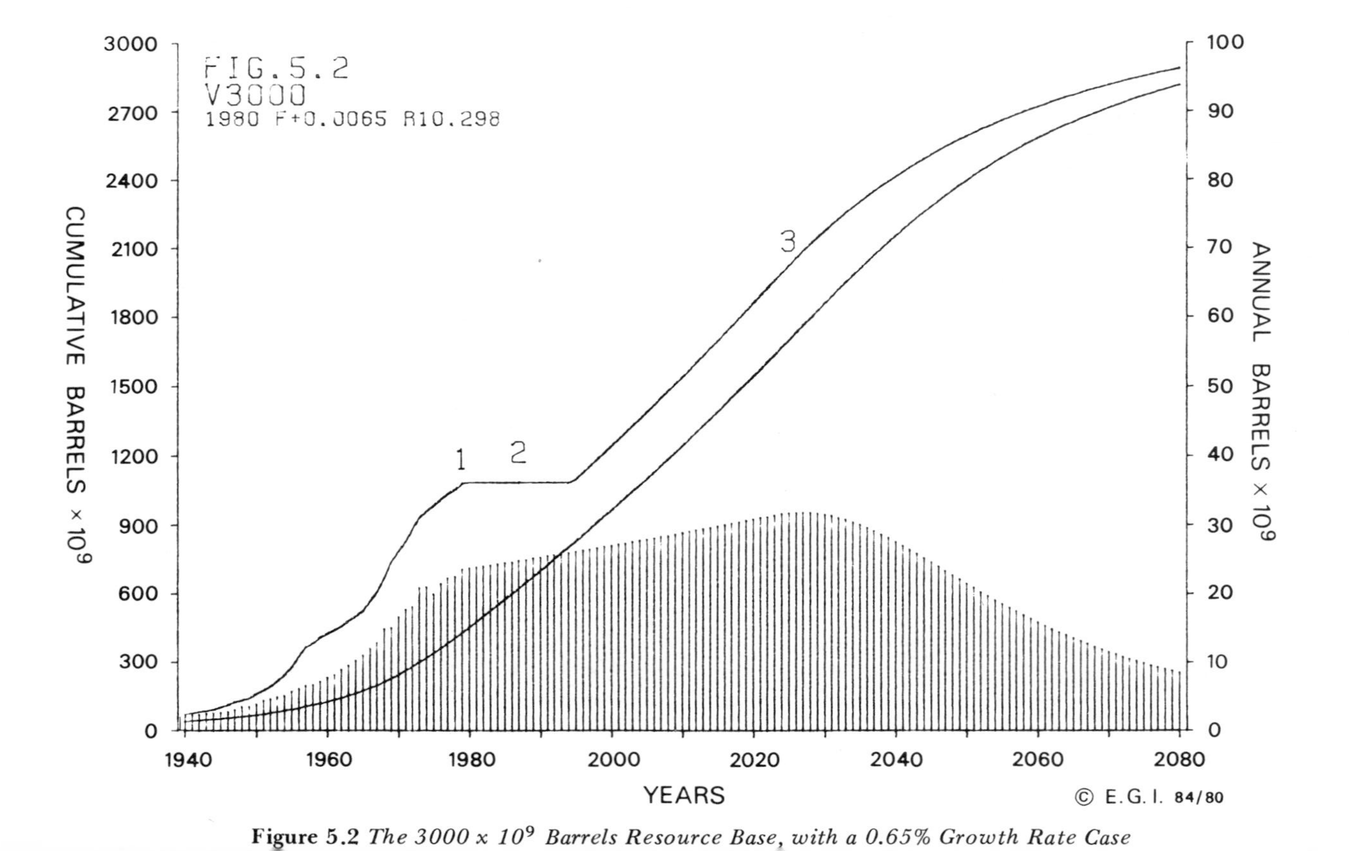

Figure 5.2 *The 3000 x 10^9 Barrels Resource Base, with a 0.65% Growth Rate Case*

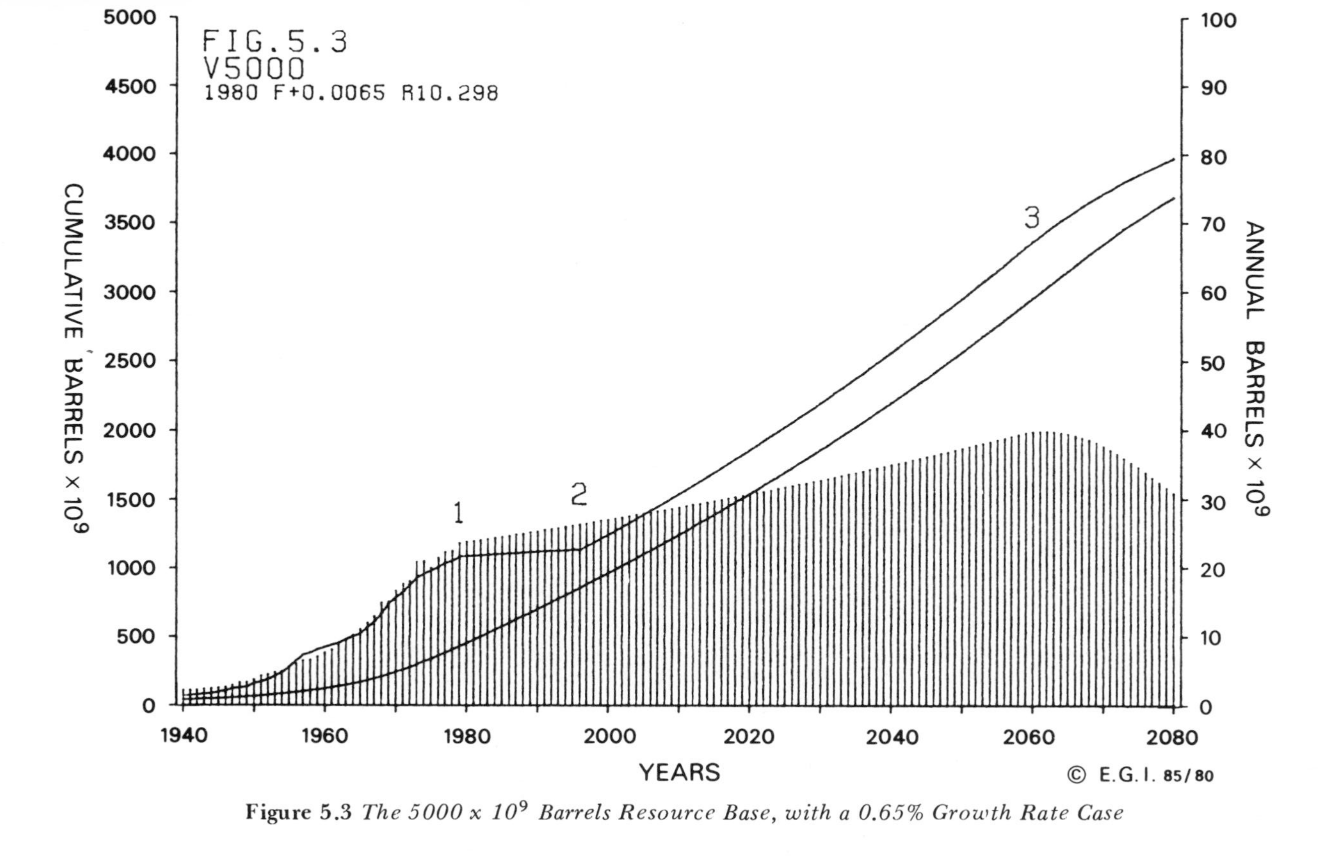

Figure 5.3 *The 5000 x 10^9 Barrels Resource Base, with a 0.65% Growth Rate Case*

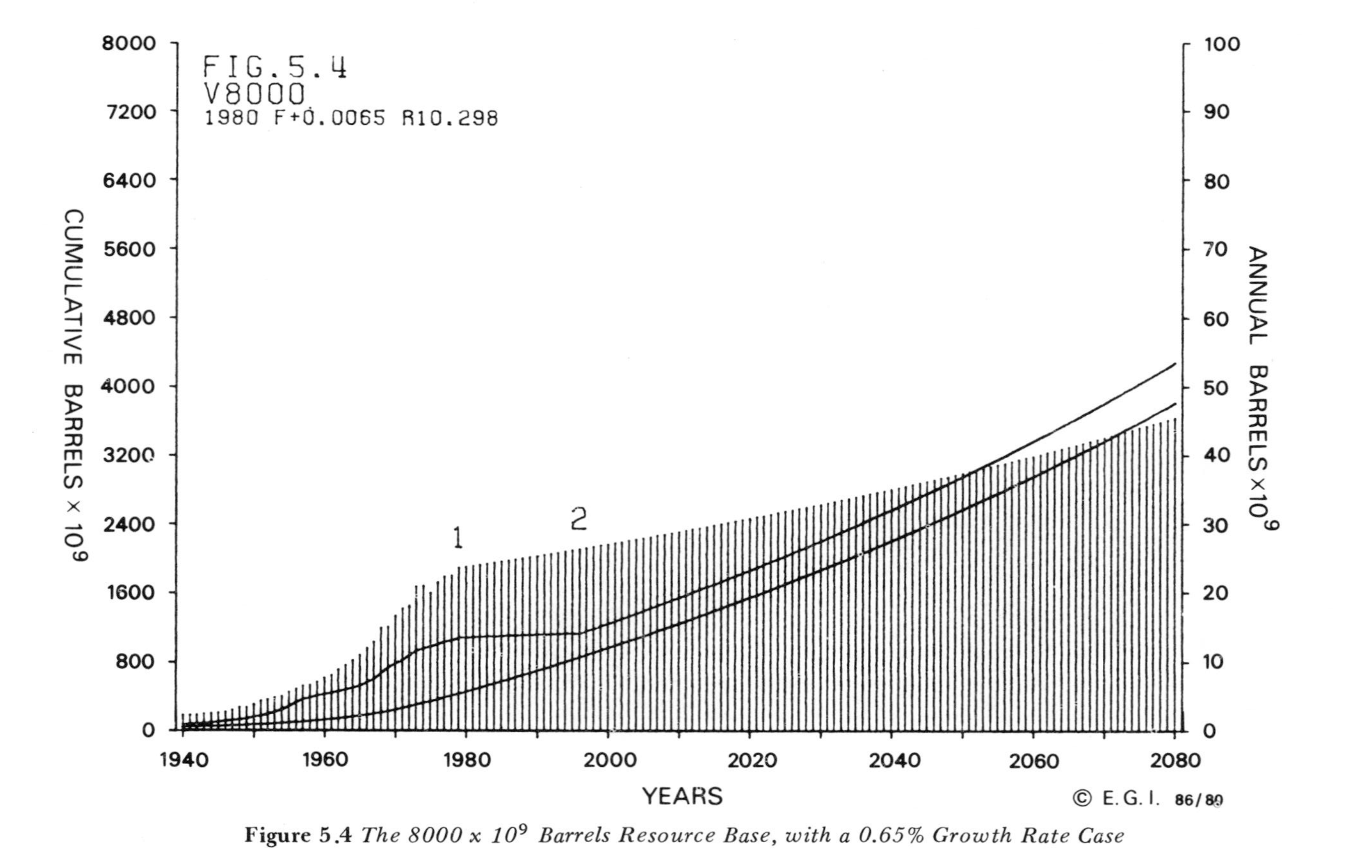

Figure 5.4 *The 8000 x 10^9 Barrels Resource Base, with a 0.65% Growth Rate Case*

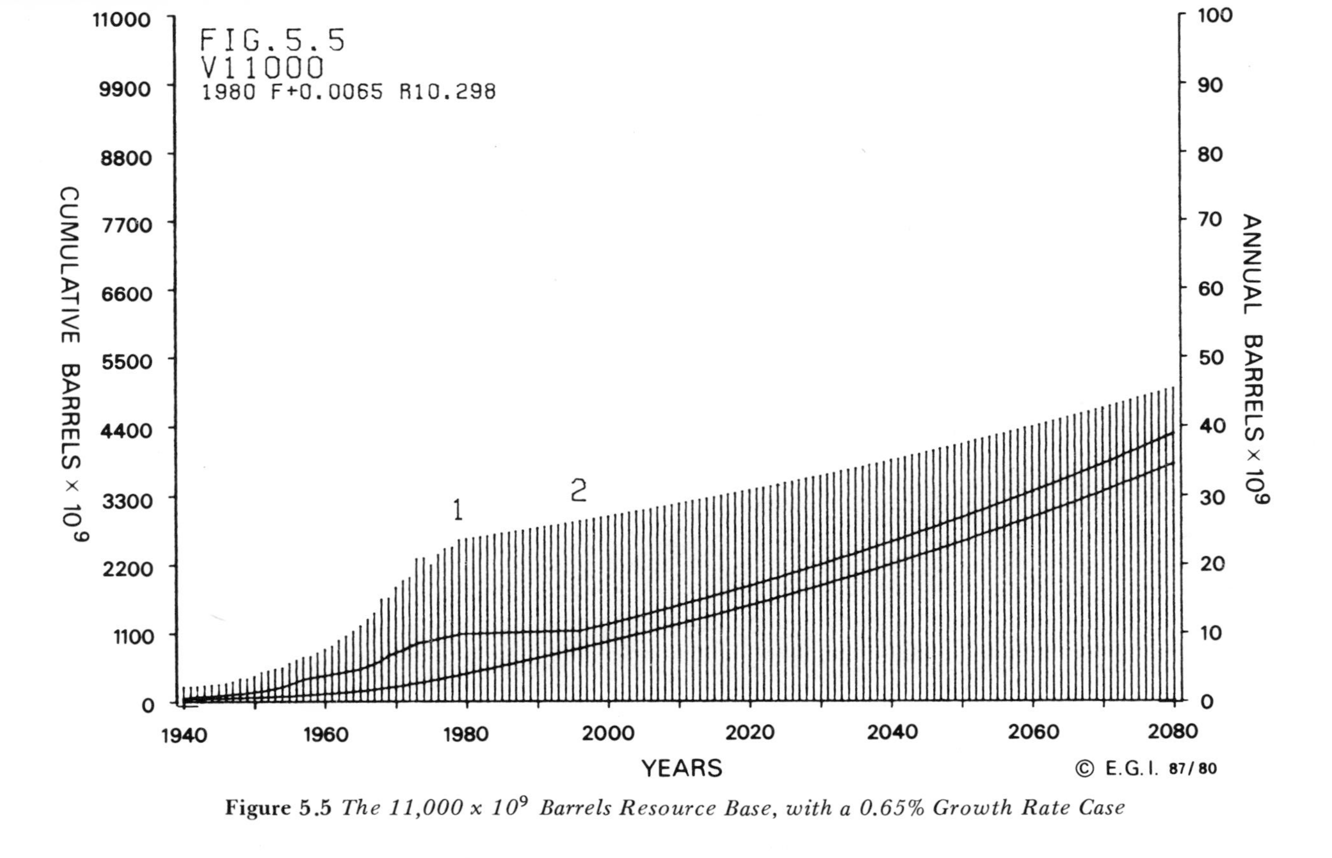

Figure 5.5 *The 11,000 x 10^9 Barrels Resource Base, with a 0.65% Growth Rate Case*

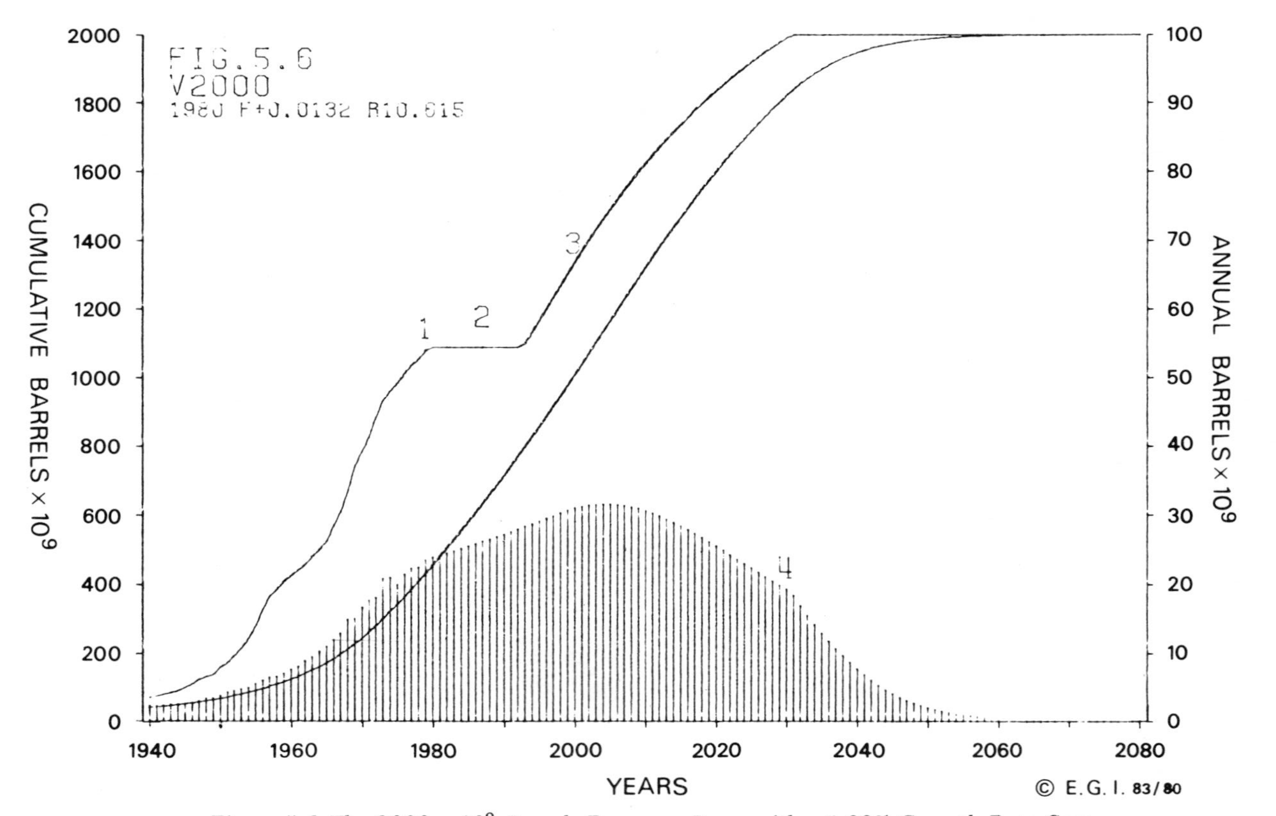

Figure 5.6 *The 2000 x 10^9 Barrels Resource Base, with a 1.32% Growth Rate Case*

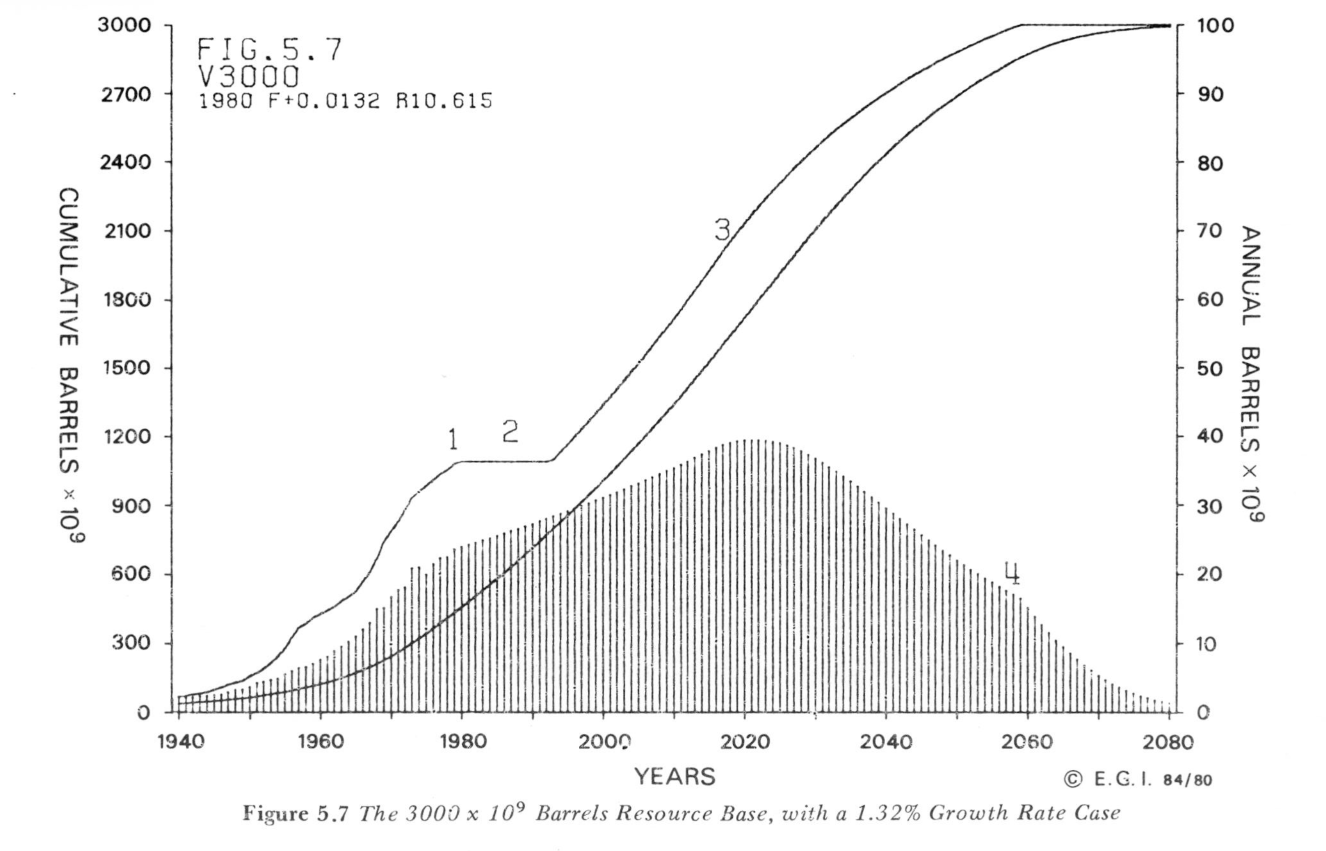

Figure 5.7 *The 3000 x 10^9 Barrels Resource Base, with a 1.32% Growth Rate Case*

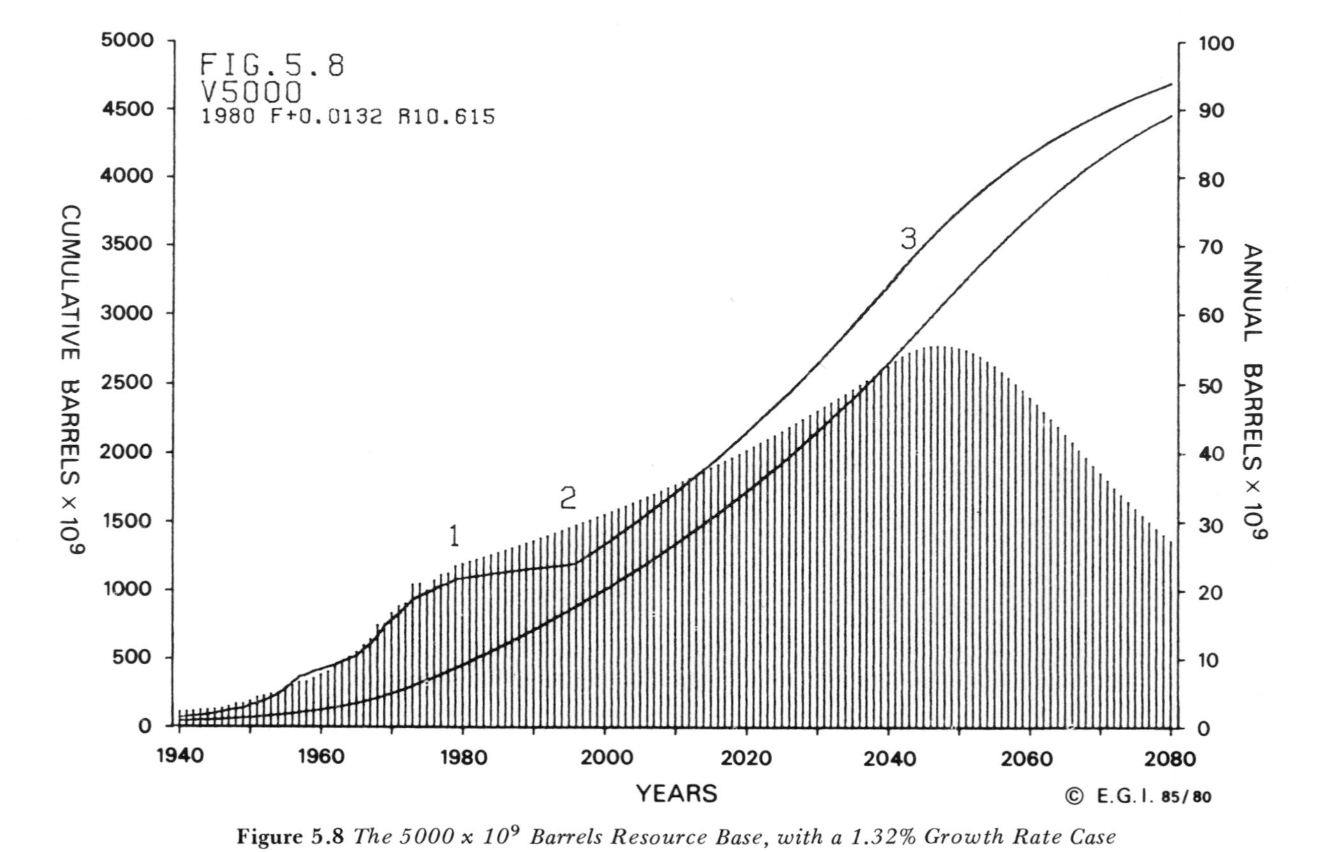

Figure 5.8 *The 5000 x 10^9 Barrels Resource Base, with a 1.32% Growth Rate Case*

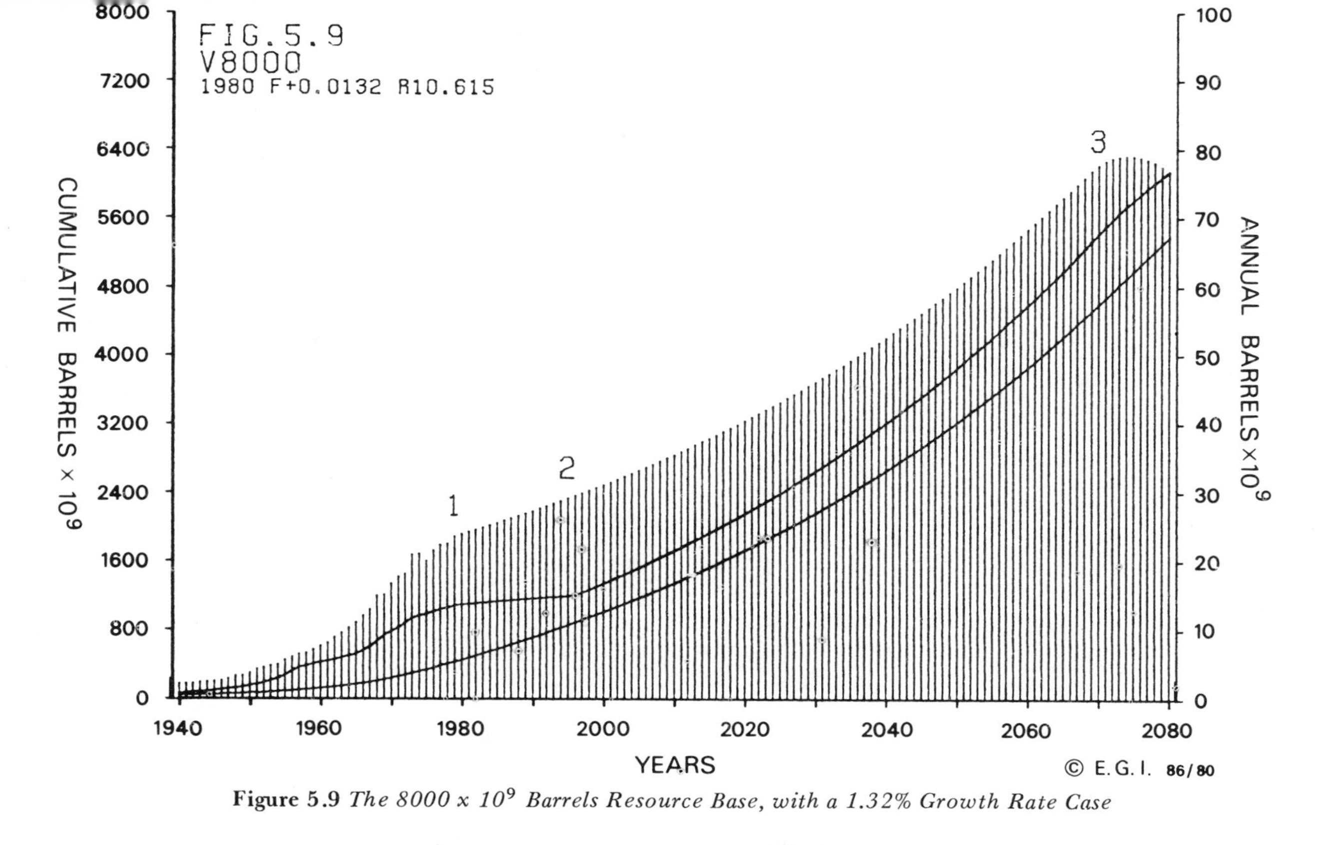

Figure 5.9 *The 8000 x 10^9 Barrels Resource Base, with a 1.32% Growth Rate Case*

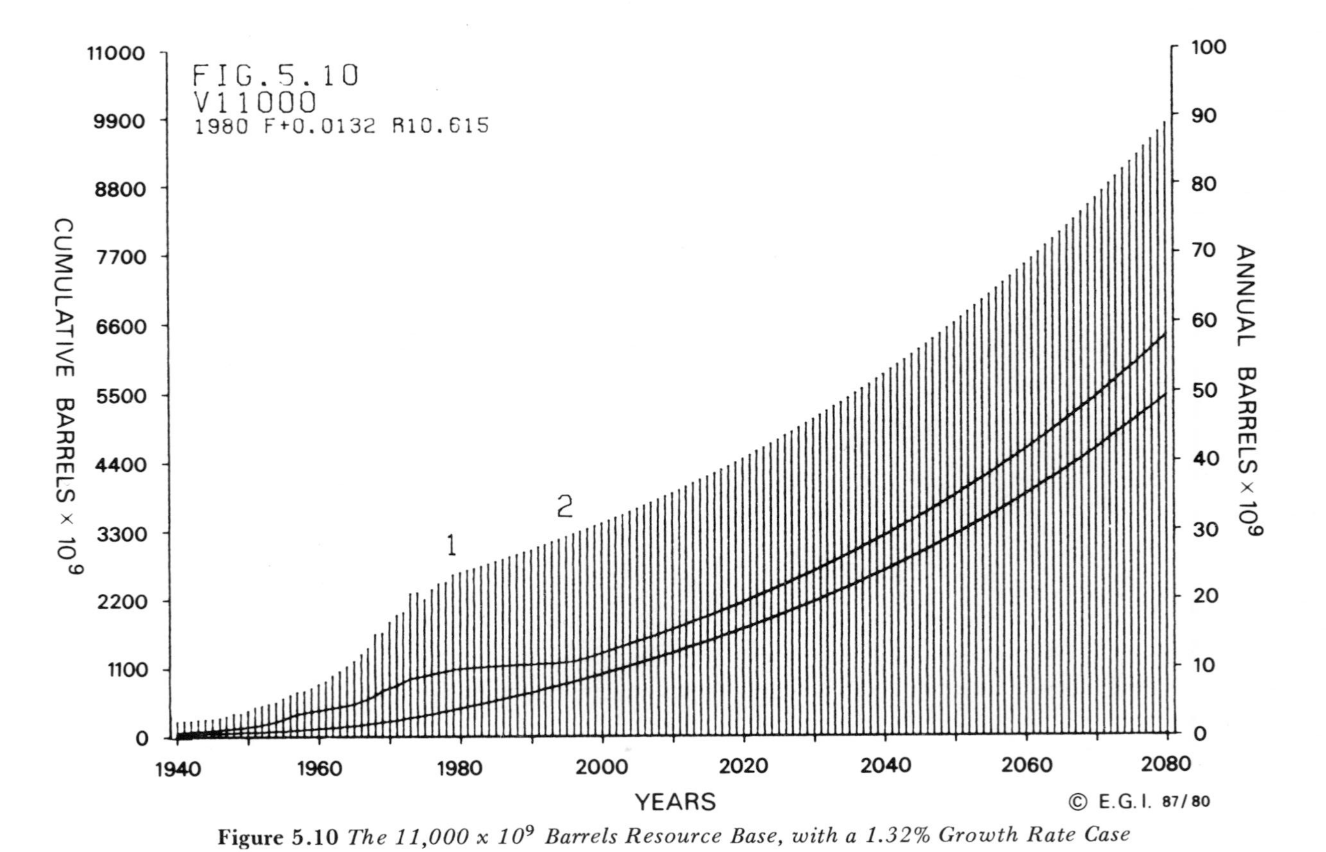

Figure 5.10 *The 11,000 x 10^9 Barrels Resource Base, with a 1.32% Growth Rate Case*

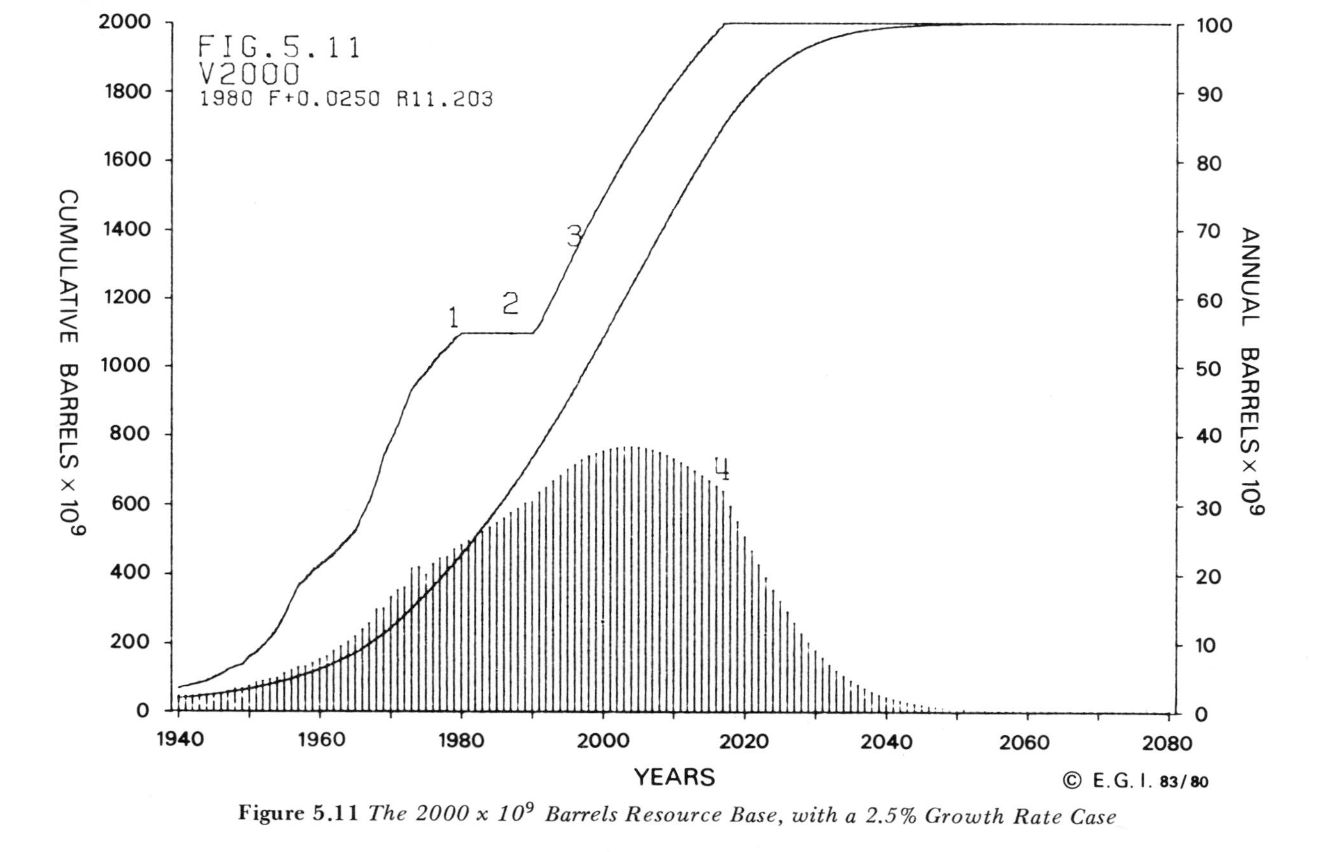

Figure 5.11 *The 2000 x 10^9 Barrels Resource Base, with a 2.5% Growth Rate Case*

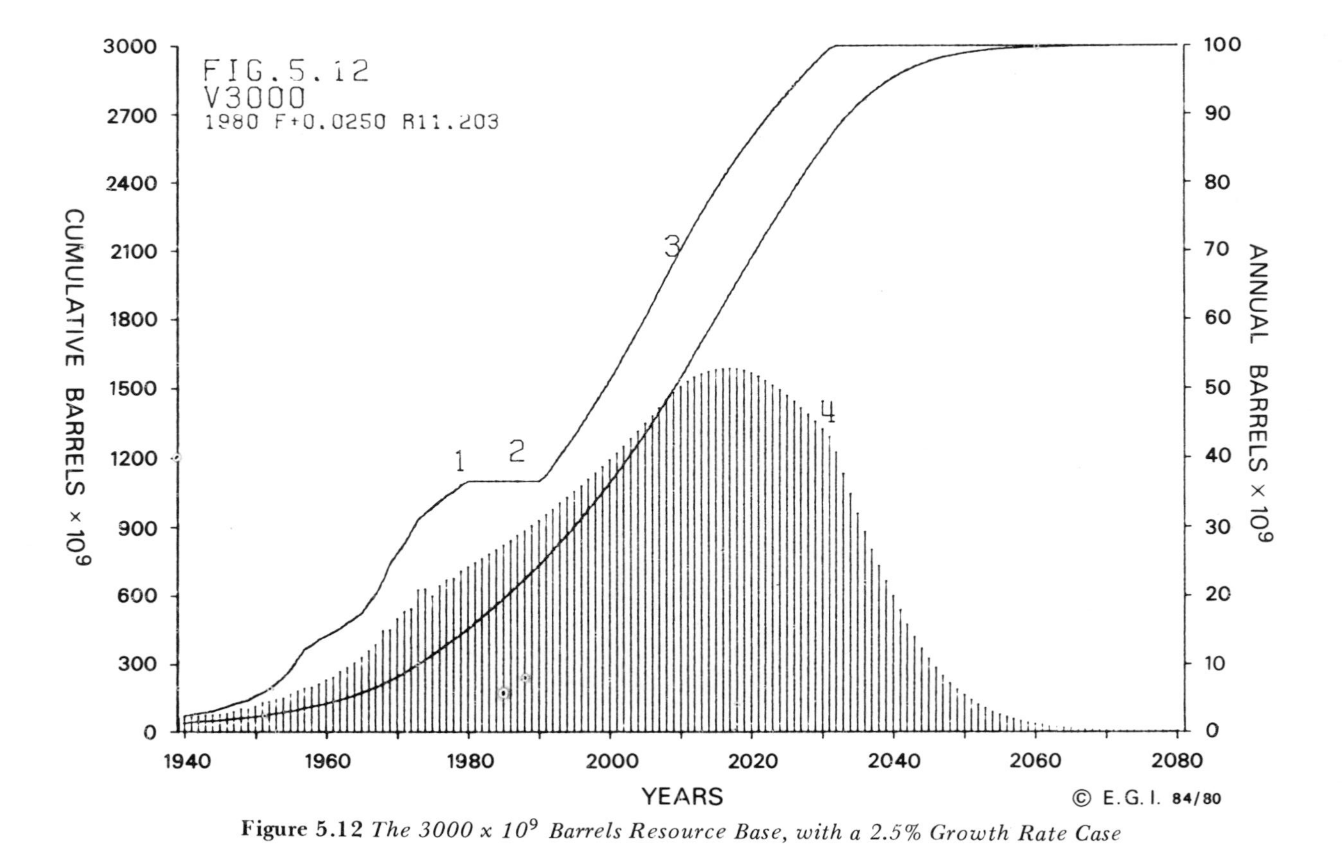

Figure 5.12 *The 3000 x 10^9 Barrels Resource Base, with a 2.5% Growth Rate Case*

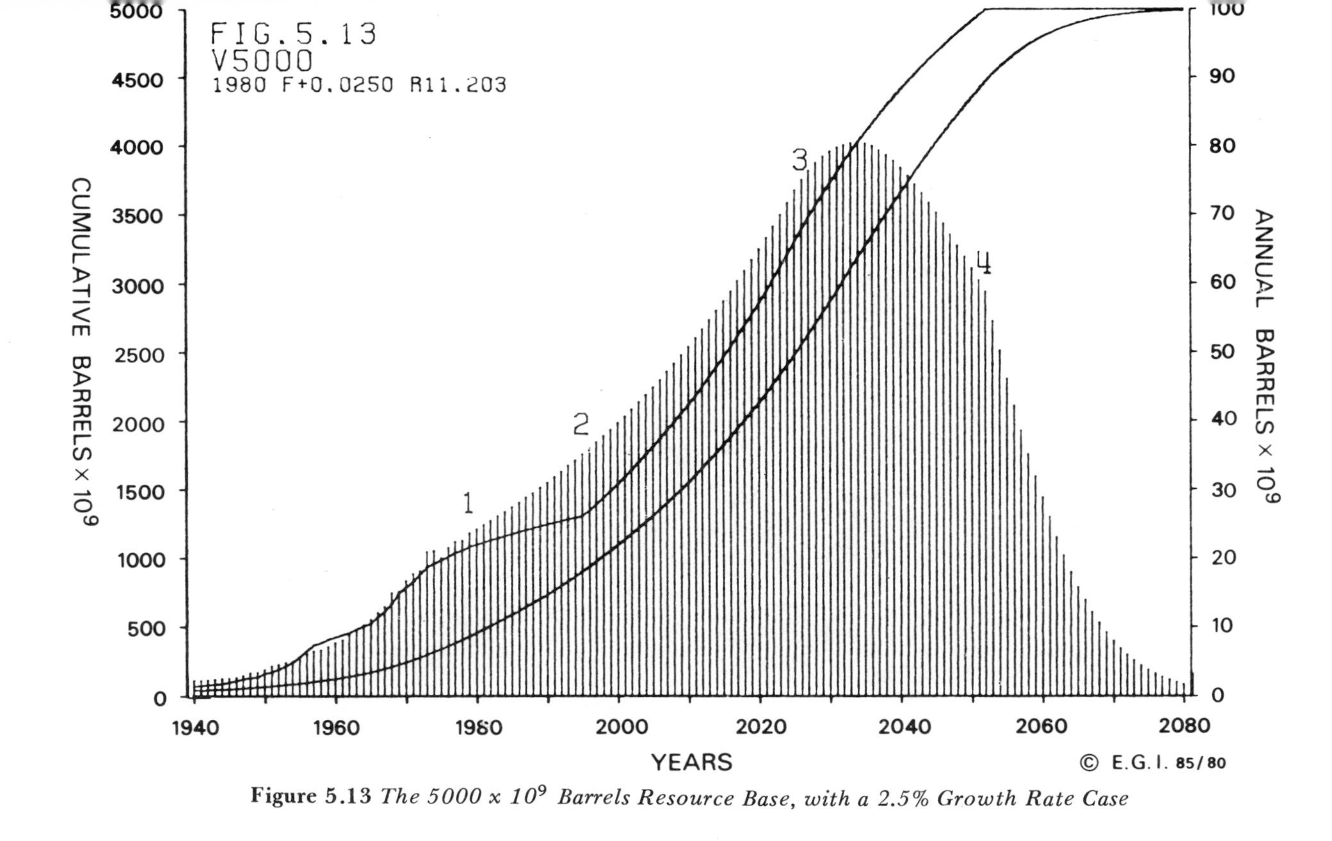

Figure 5.13 *The 5000 x 10^9 Barrels Resource Base, with a 2.5% Growth Rate Case*

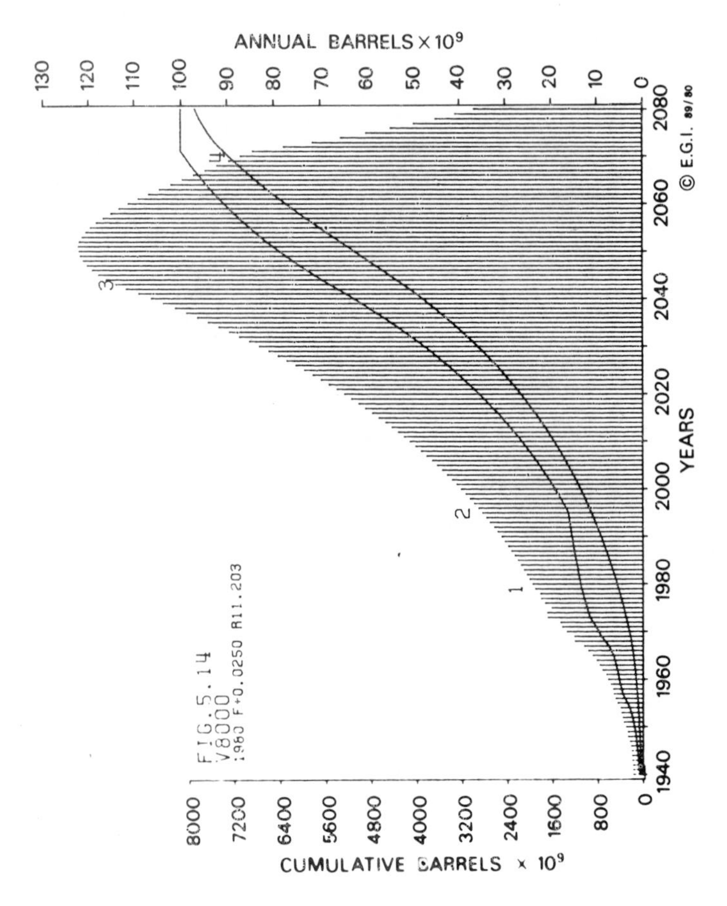

Figure 5.14 *The 8000 x 10⁹ Barrels Resource Base, with a 2.5% Growth Rate Case*

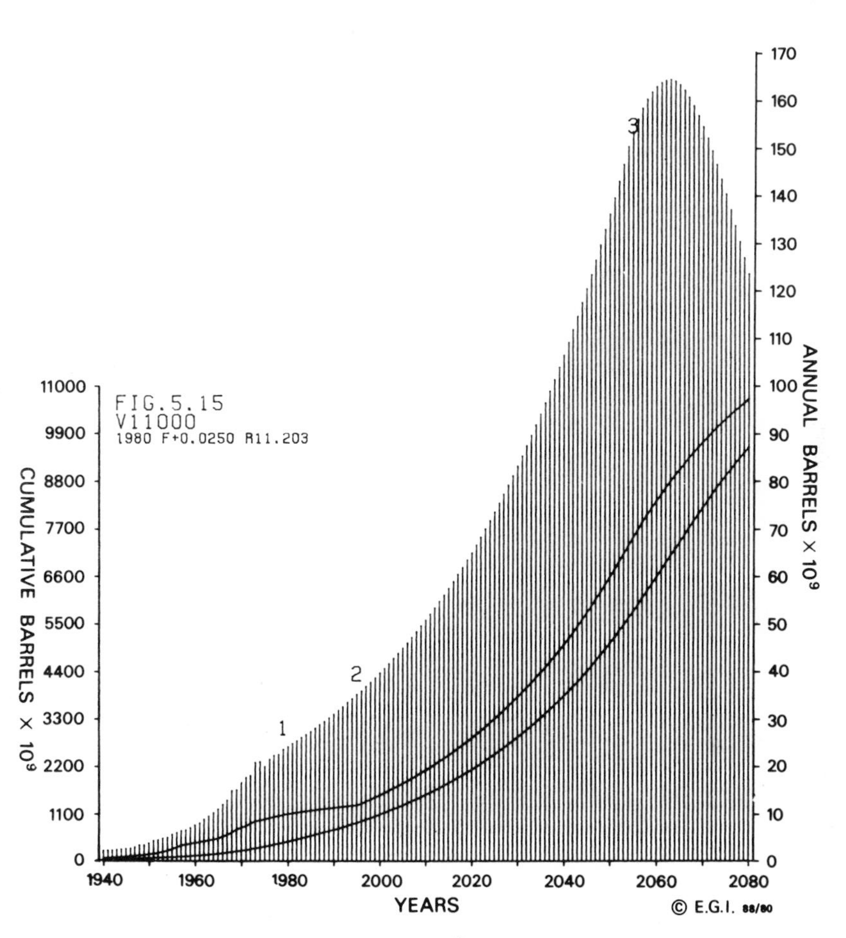

Figure 5.15 *The 11,000 x 10^9 Barrels Resource Base, with a 2.5% Growth Rate Case*

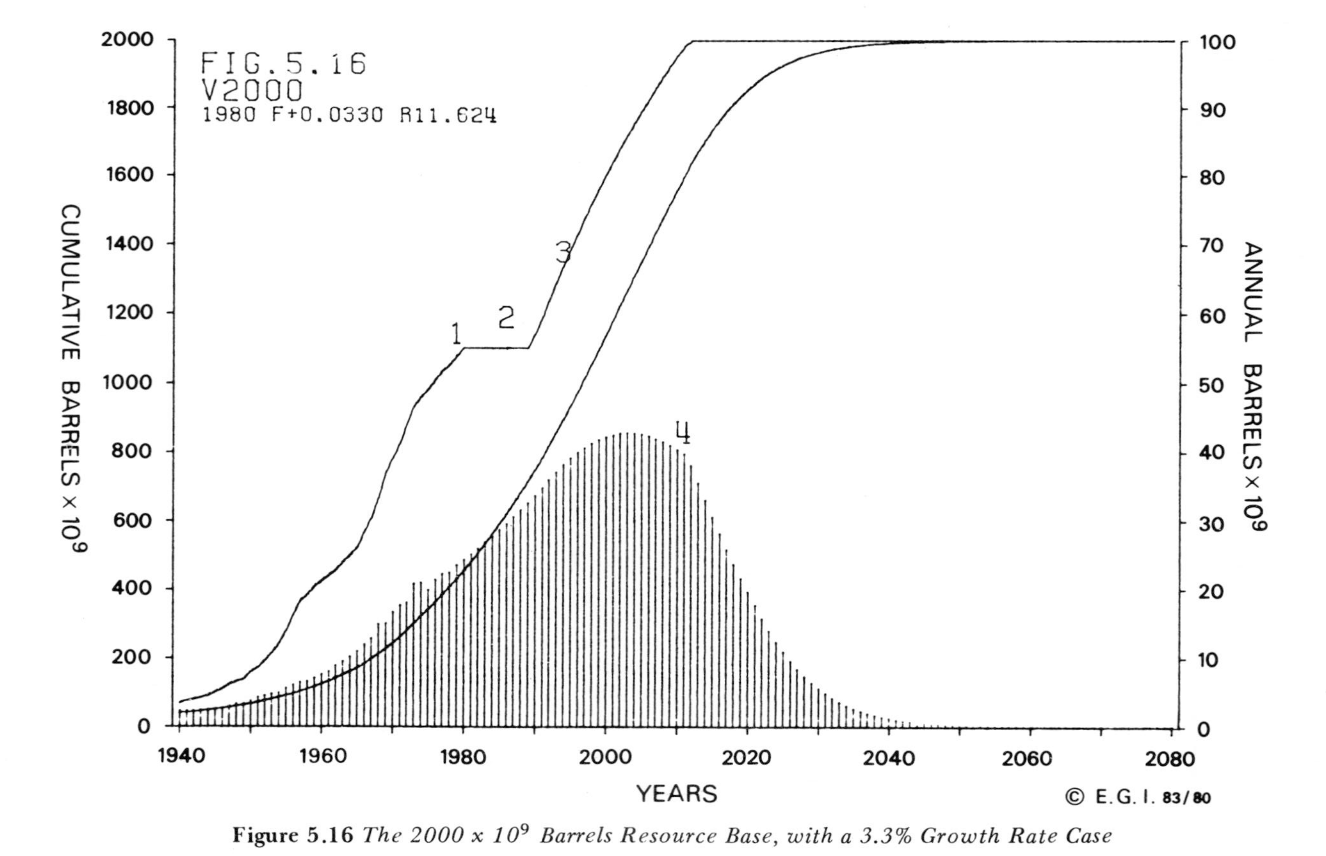

Figure 5.16 *The 2000 x 10^9 Barrels Resource Base, with a 3.3% Growth Rate Case*

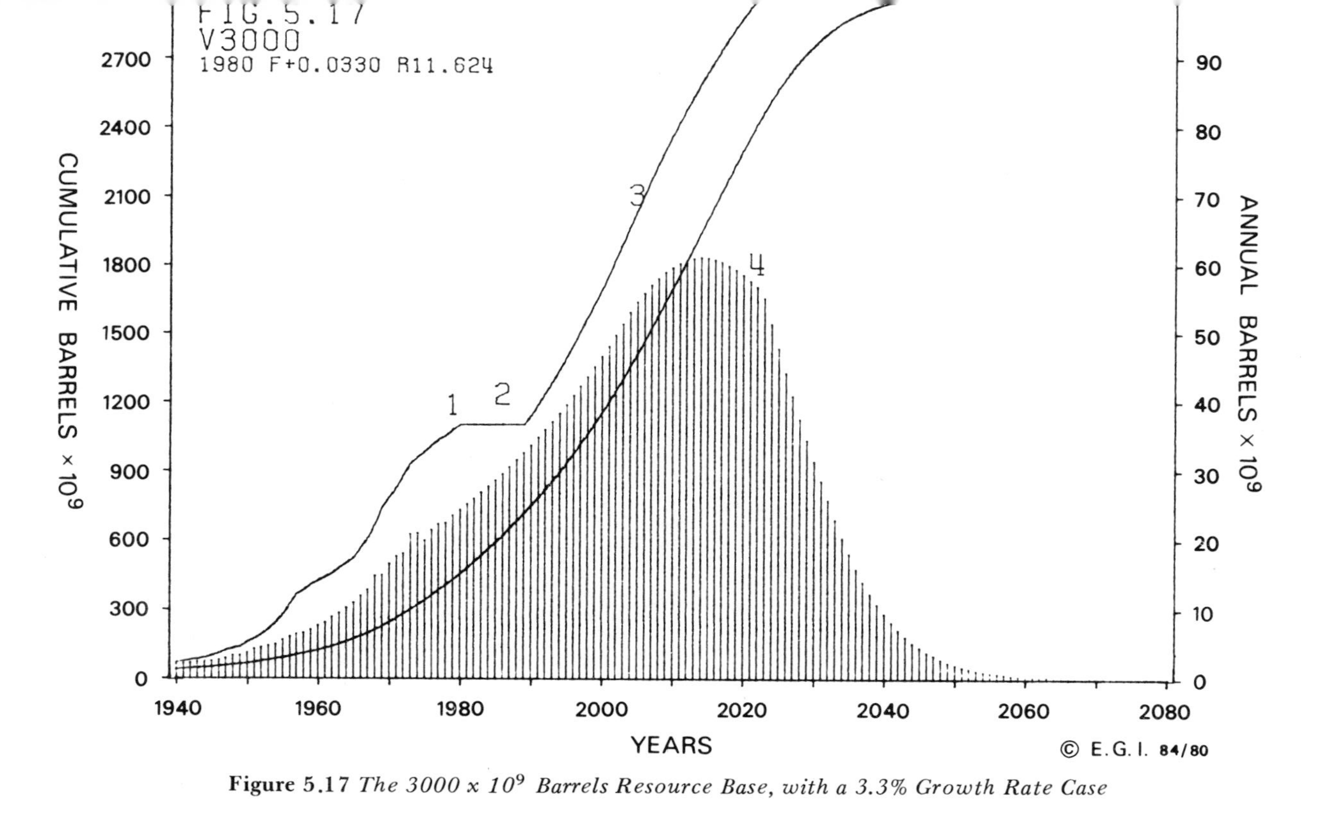

Figure 5.17 *The 3000 x 10^9 Barrels Resource Base, with a 3.3% Growth Rate Case*

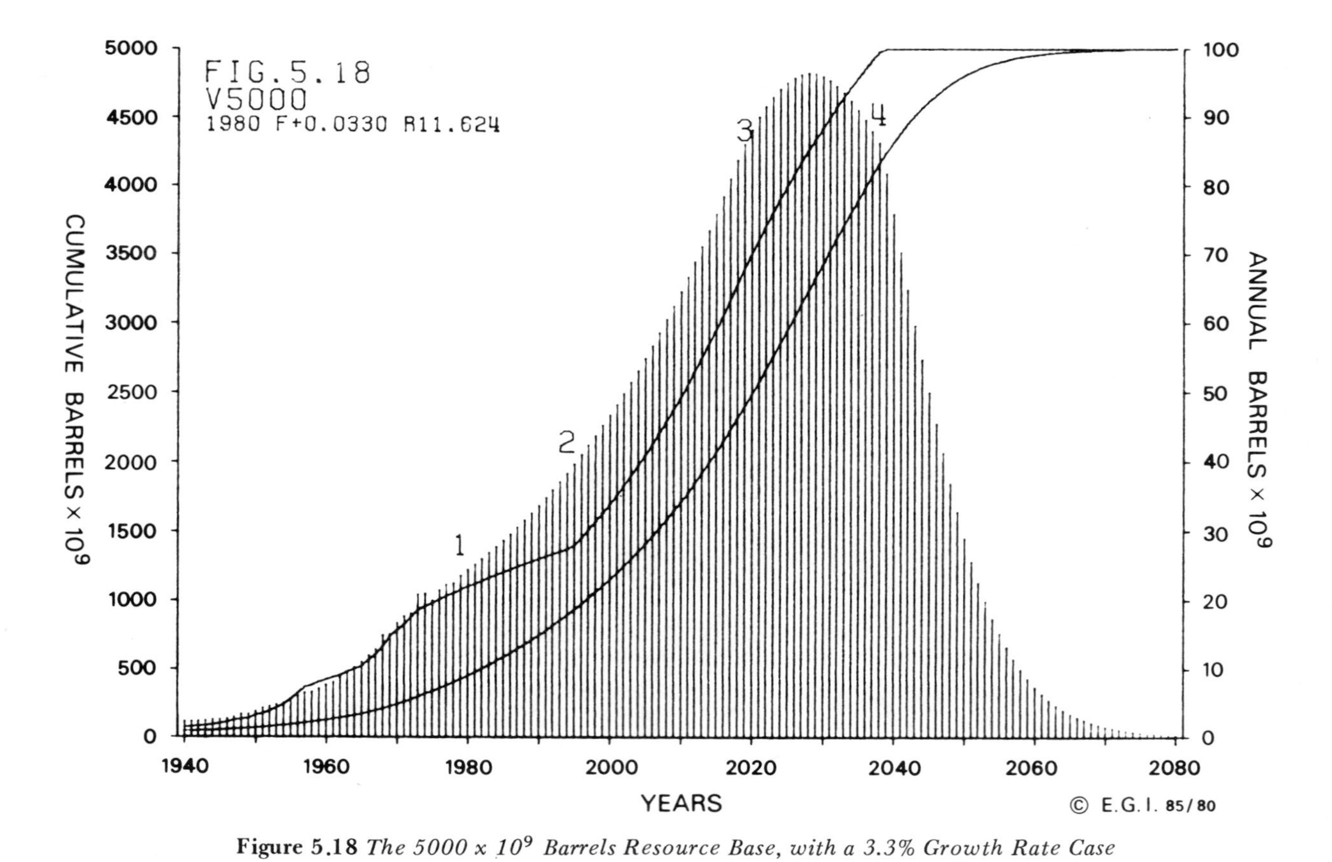

Figure 5.18 *The 5000 x 10^9 Barrels Resource Base, with a 3.3% Growth Rate Case*

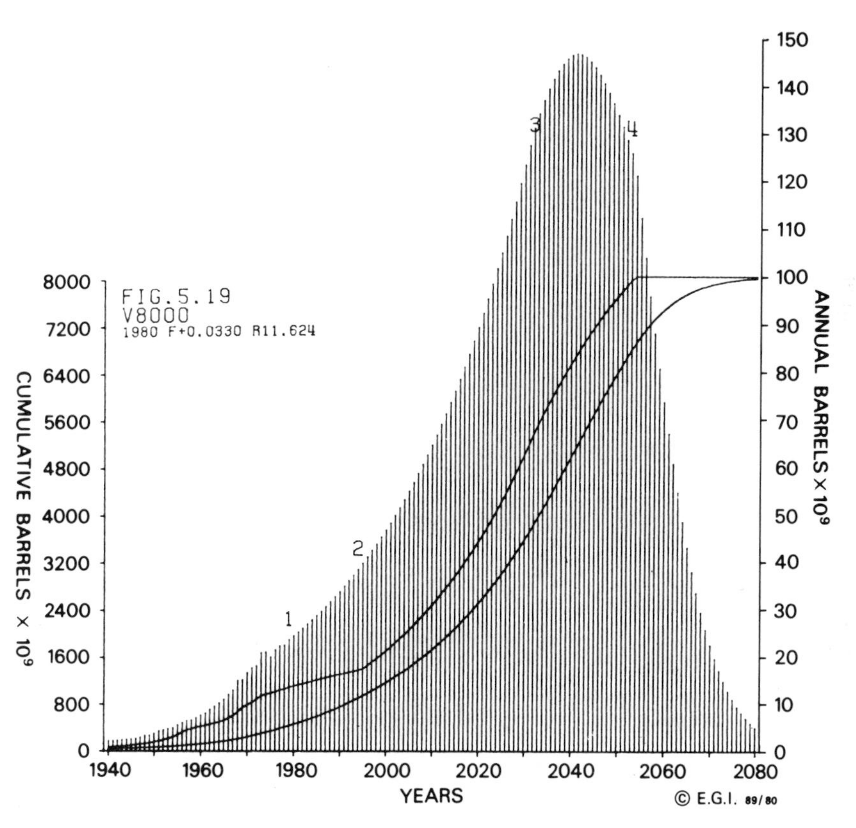

Figure 5.19 *The 8000 x 10^9 Barrels Resource Base, with a 3.3% Growth Rate Case*

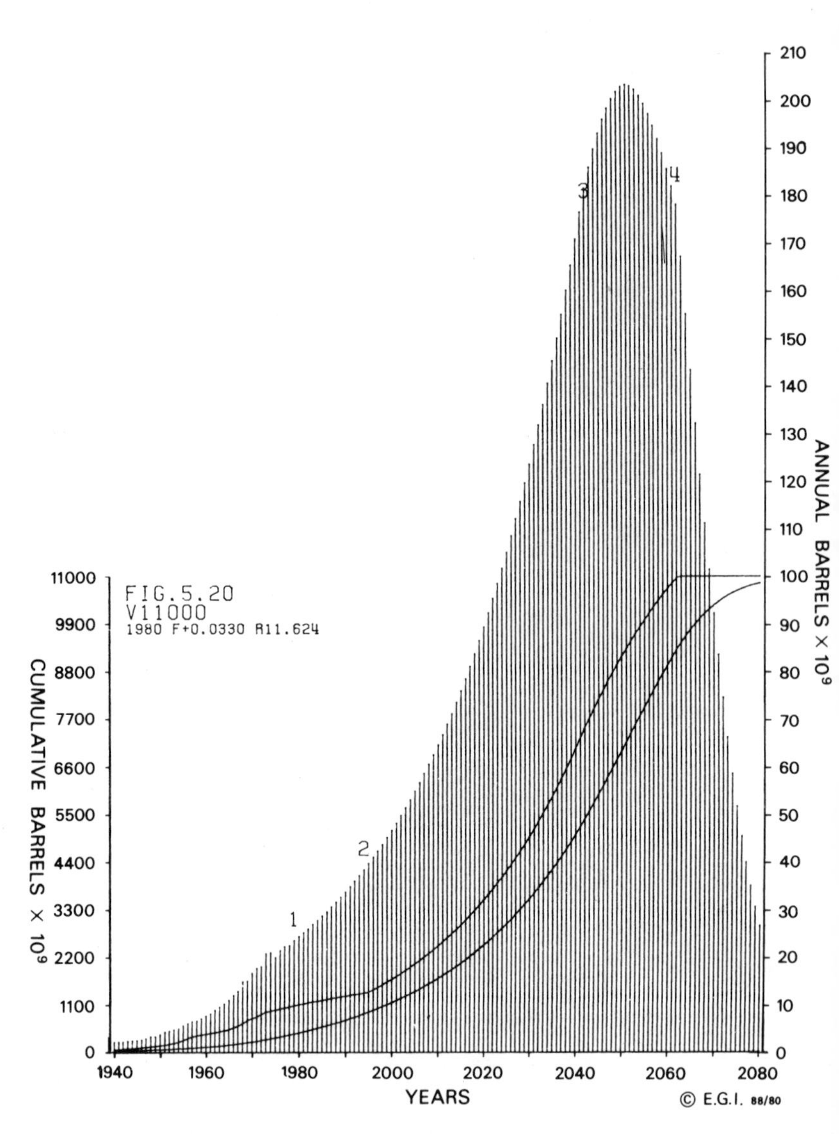

Figure 5.20 *The 11,000 x 10^9 Barrels Resource Base, with a 3.3% Growth Rate Case*

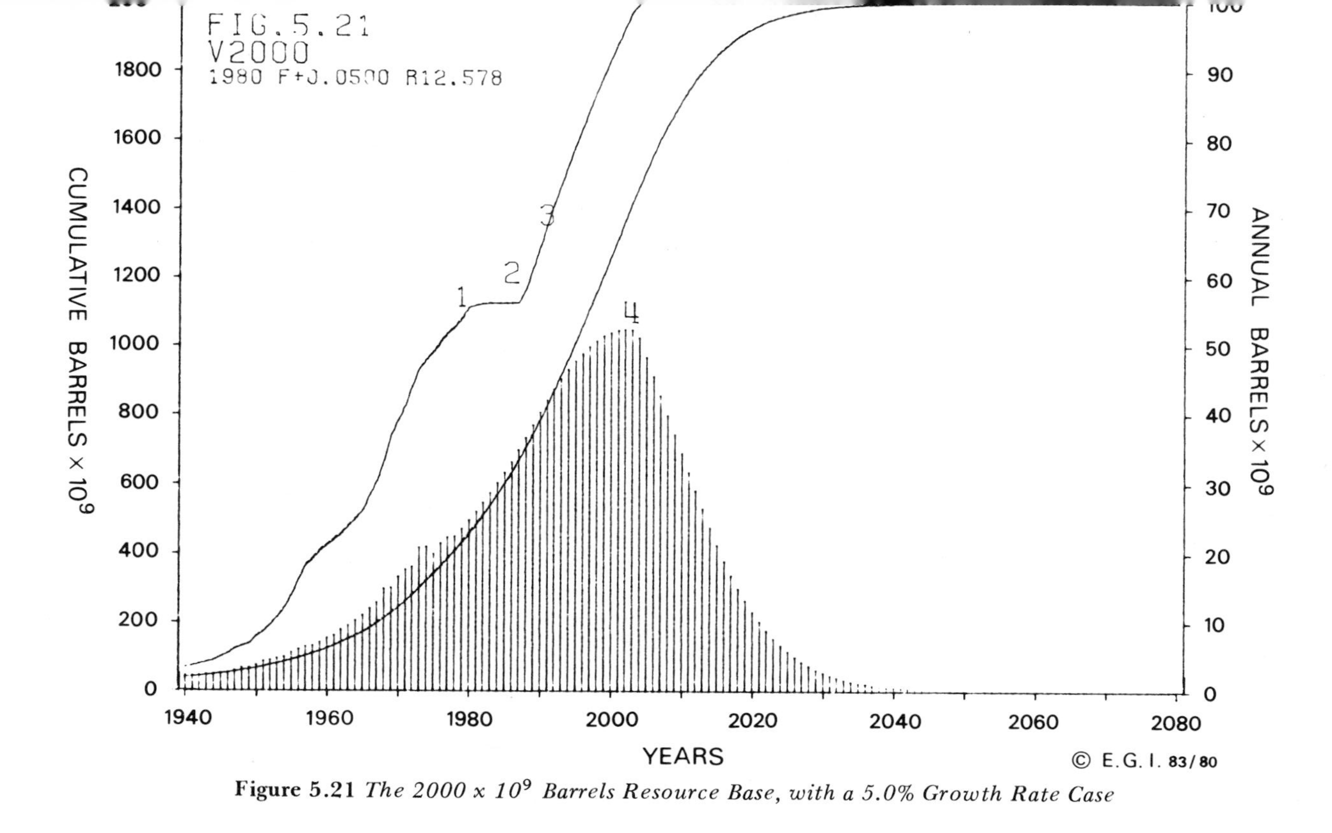

Figure 5.21 *The 2000 x 10^9 Barrels Resource Base, with a 5.0% Growth Rate Case*

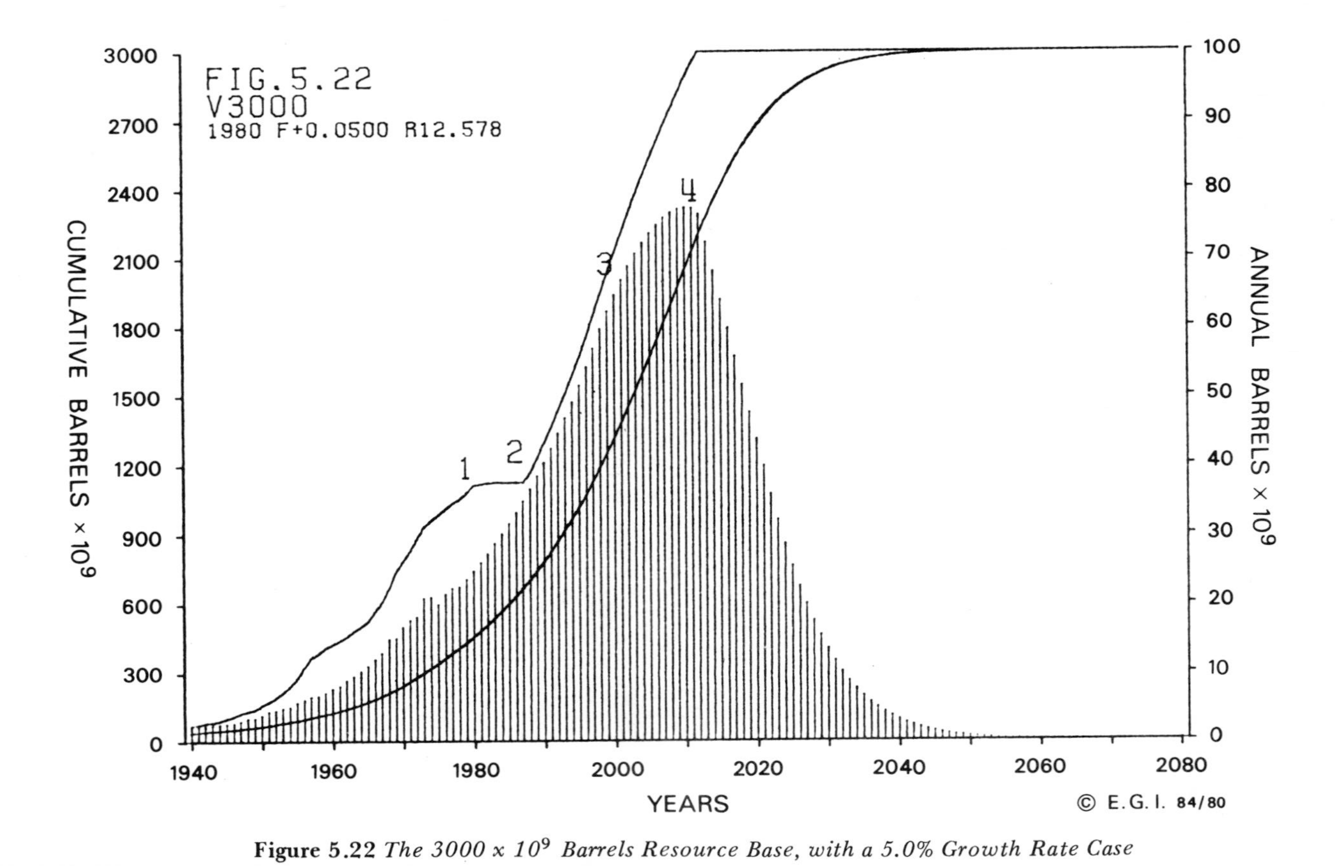

Figure 5.22 *The 3000 x 10^9 Barrels Resource Base, with a 5.0% Growth Rate Case*

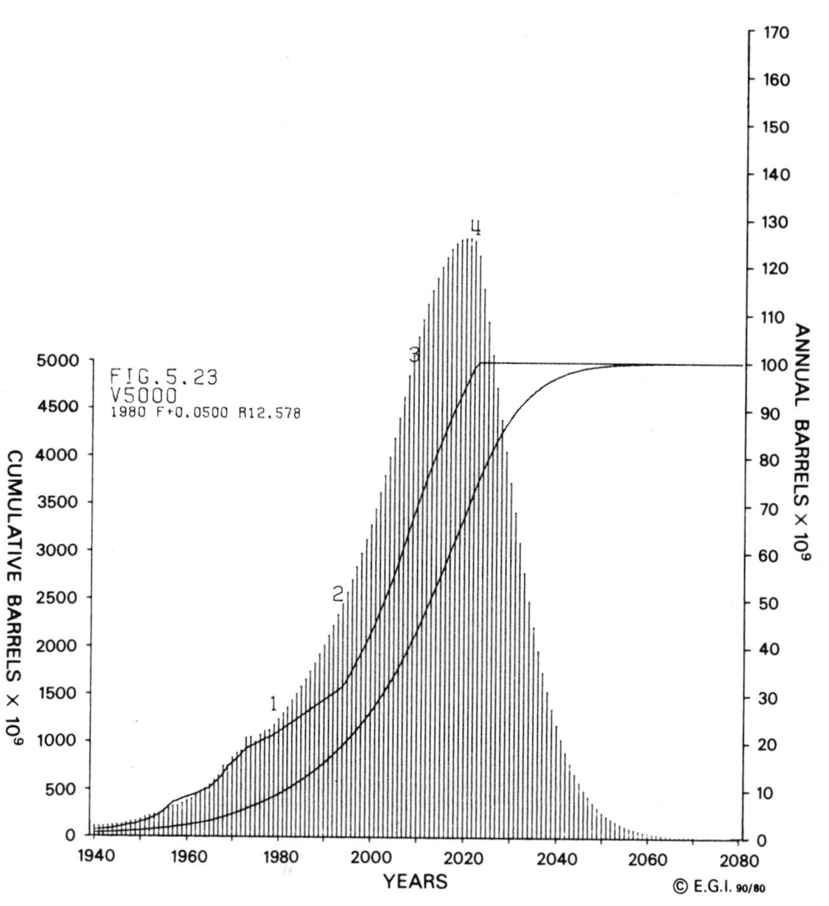

Figure 5.23 *The 5000 x 10^9 Barrels Resource Base, with a 5.0% Growth Rate Case*

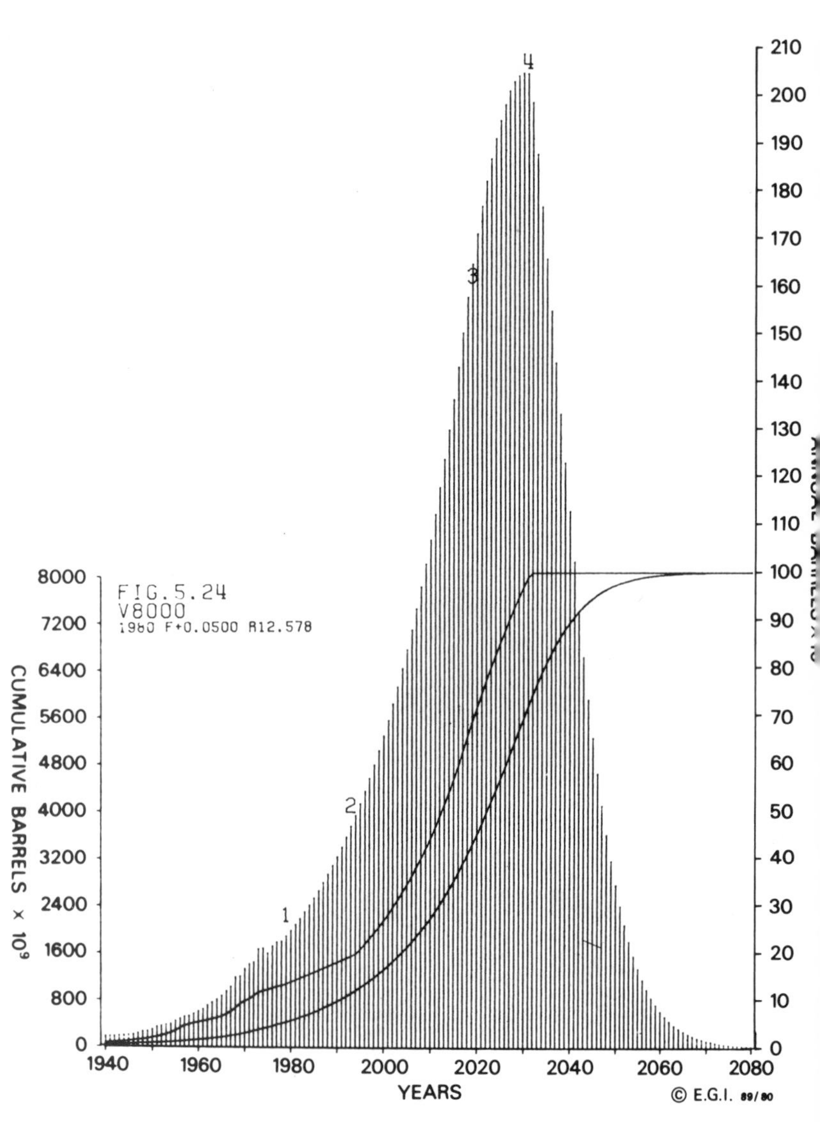

Figure 5.24 *The 8000 x 10^9 Barrels Resource Base, with a 5.0% Growth Rate Case*

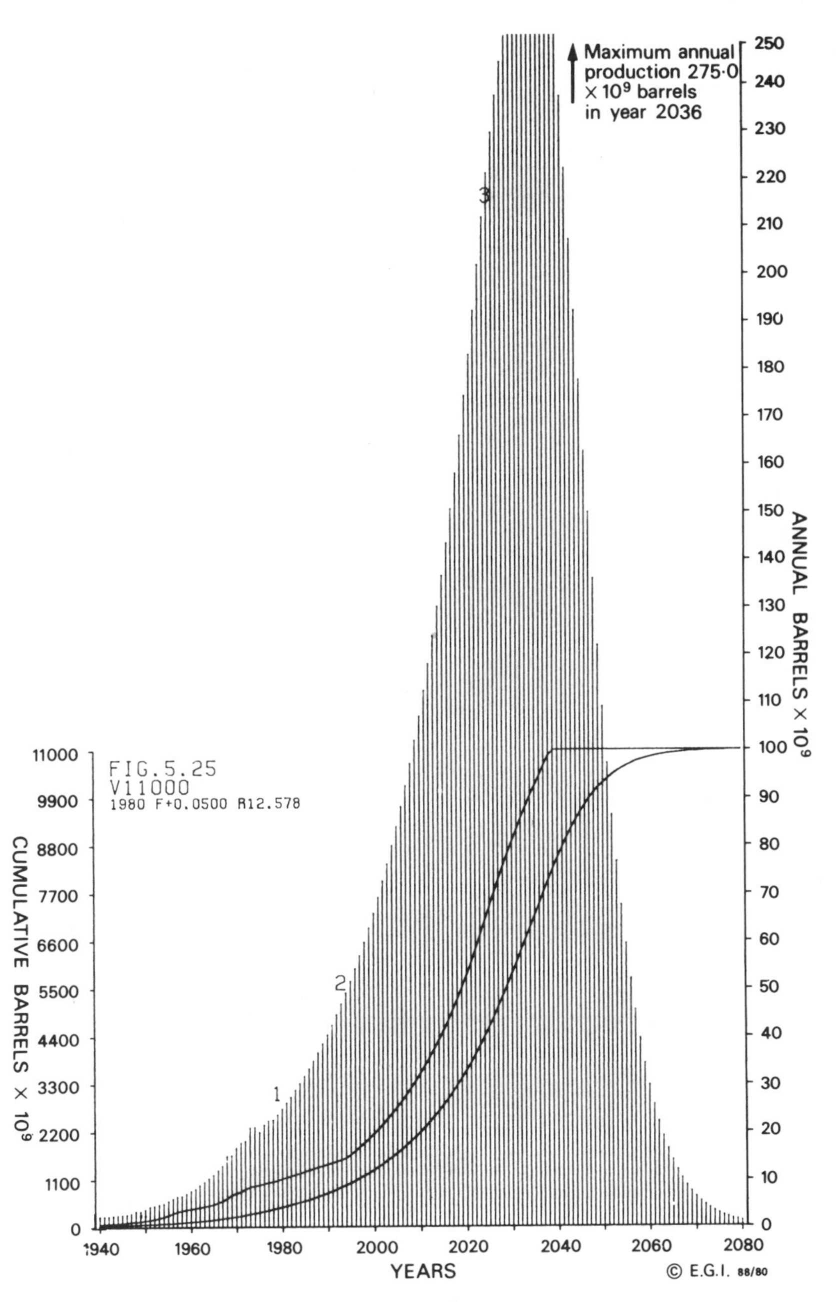

Figure 5.25 *The 11,000 x 10^9 Barrels Resource Base, with a 5.0% Growth Rate Case*

Appendix 5.2

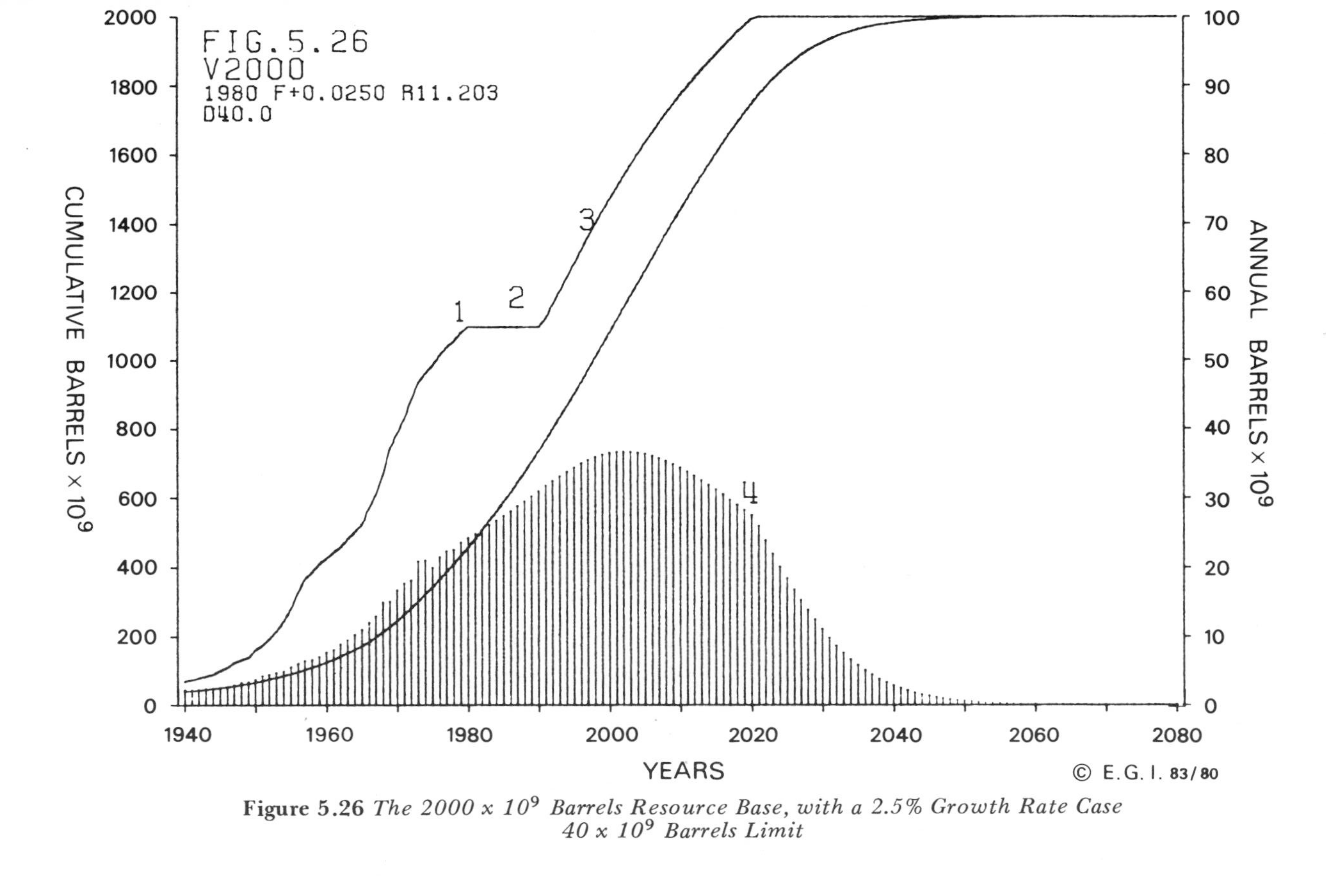

Figure 5.26 *The 2000 x 10^9 Barrels Resource Base, with a 2.5% Growth Rate Case 40 x 10^9 Barrels Limit*

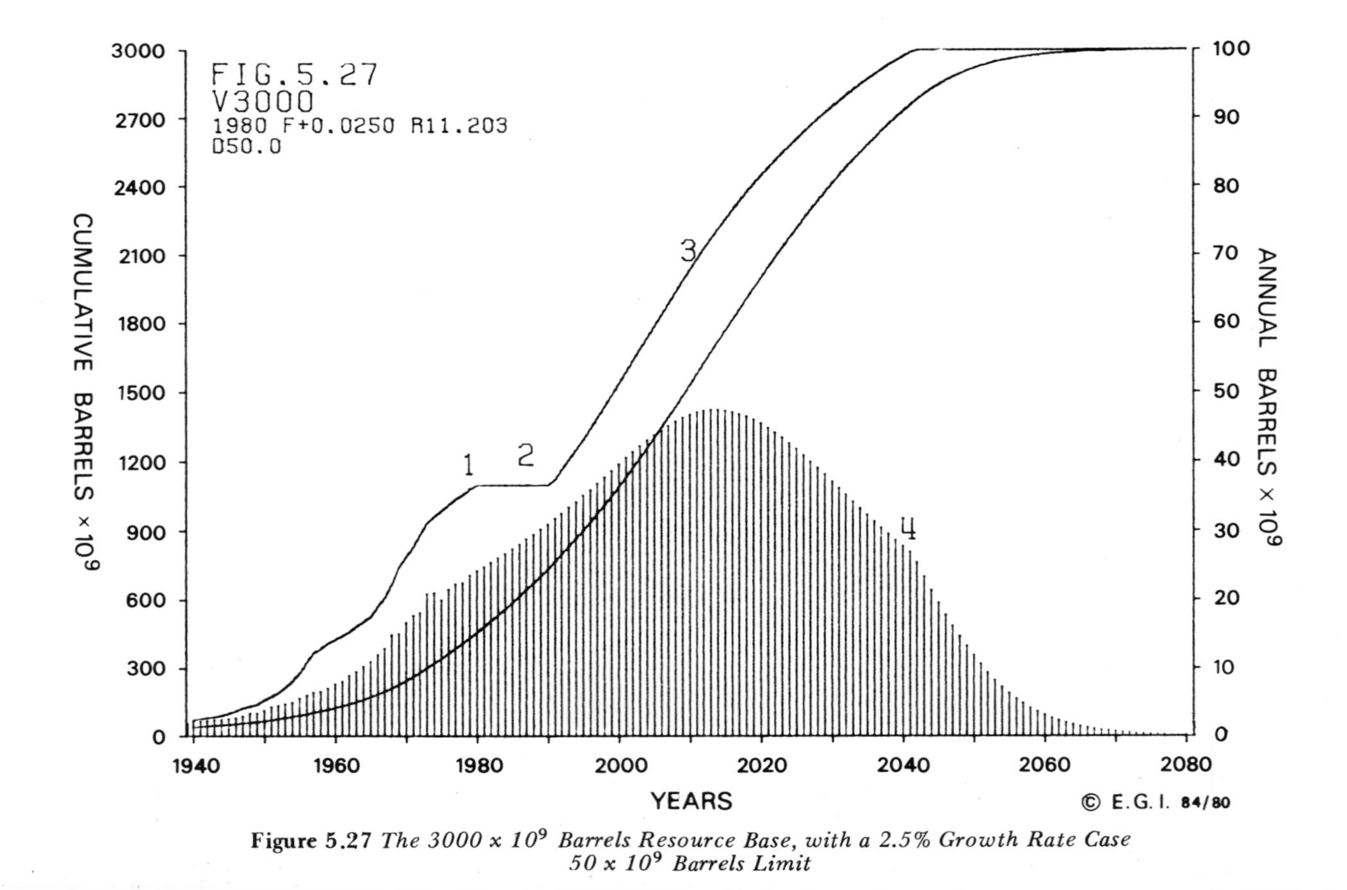

Figure 5.27 *The 3000 x 10^9 Barrels Resource Base, with a 2.5% Growth Rate Case 50 x 10^9 Barrels Limit*

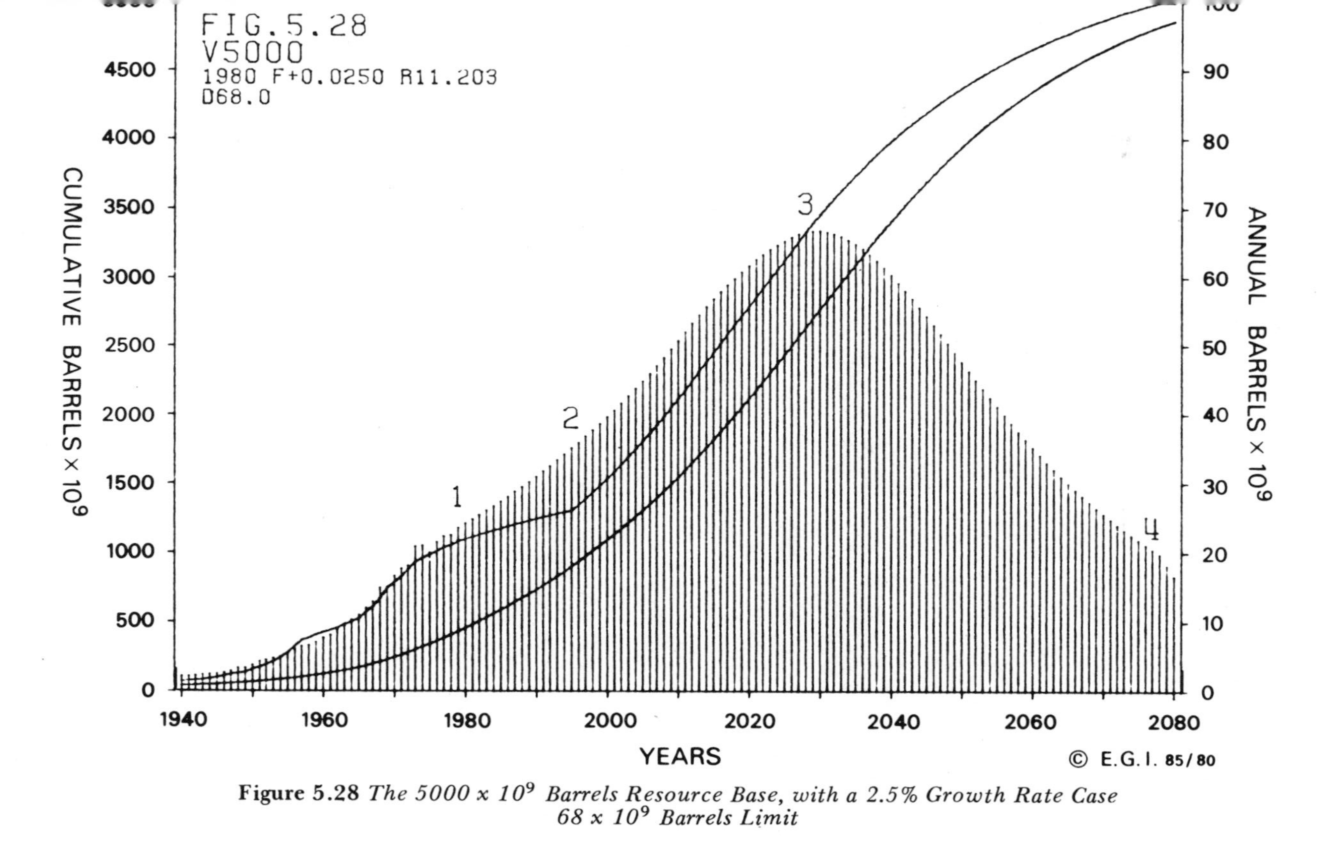

Figure 5.28 *The 5000 x 10^9 Barrels Resource Base, with a 2.5% Growth Rate Case 68 x 10^9 Barrels Limit*

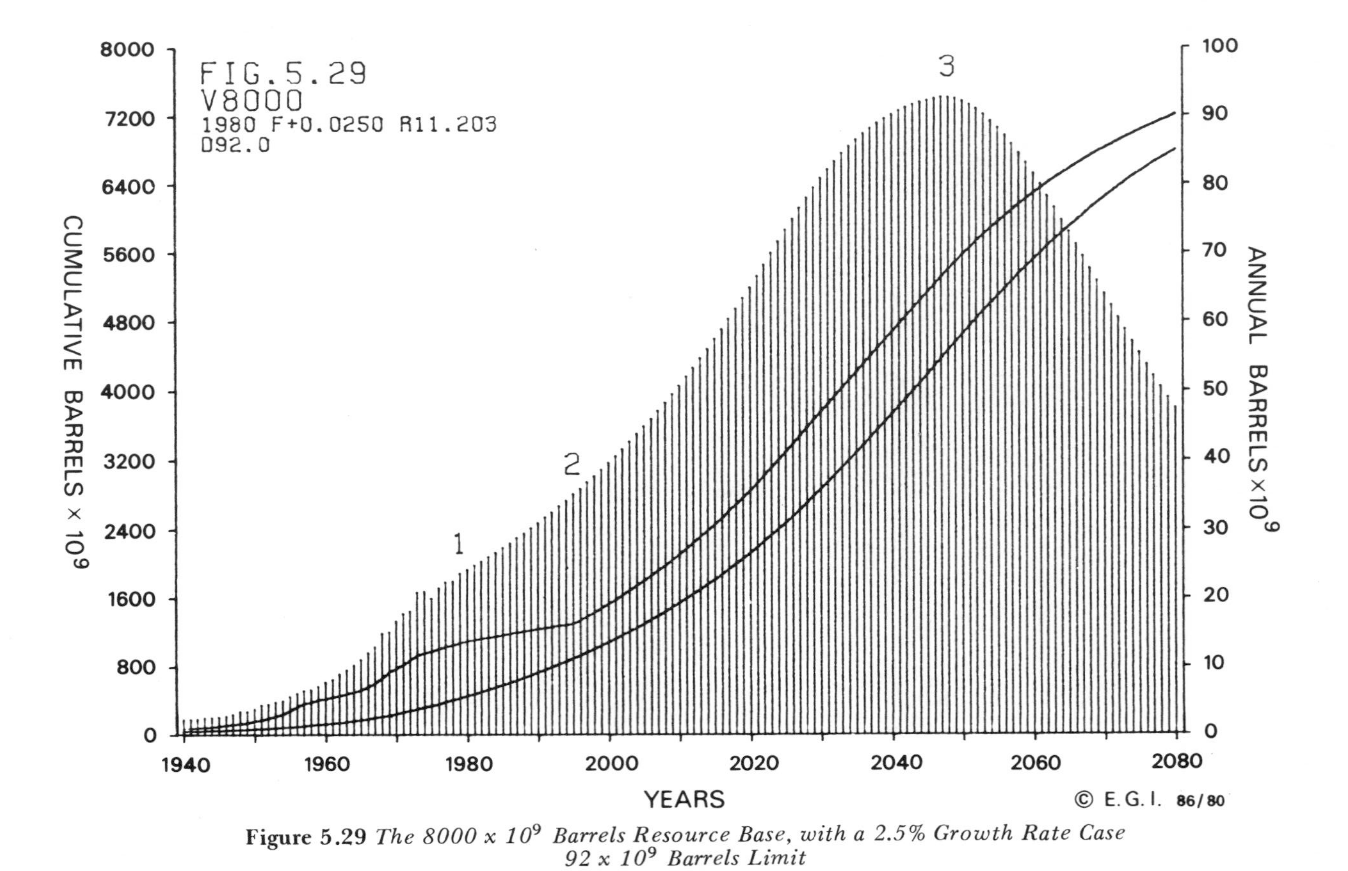

Figure 5.29 *The 8000 x 10^9 Barrels Resource Base, with a 2.5% Growth Rate Case*
92 x 10^9 Barrels Limit

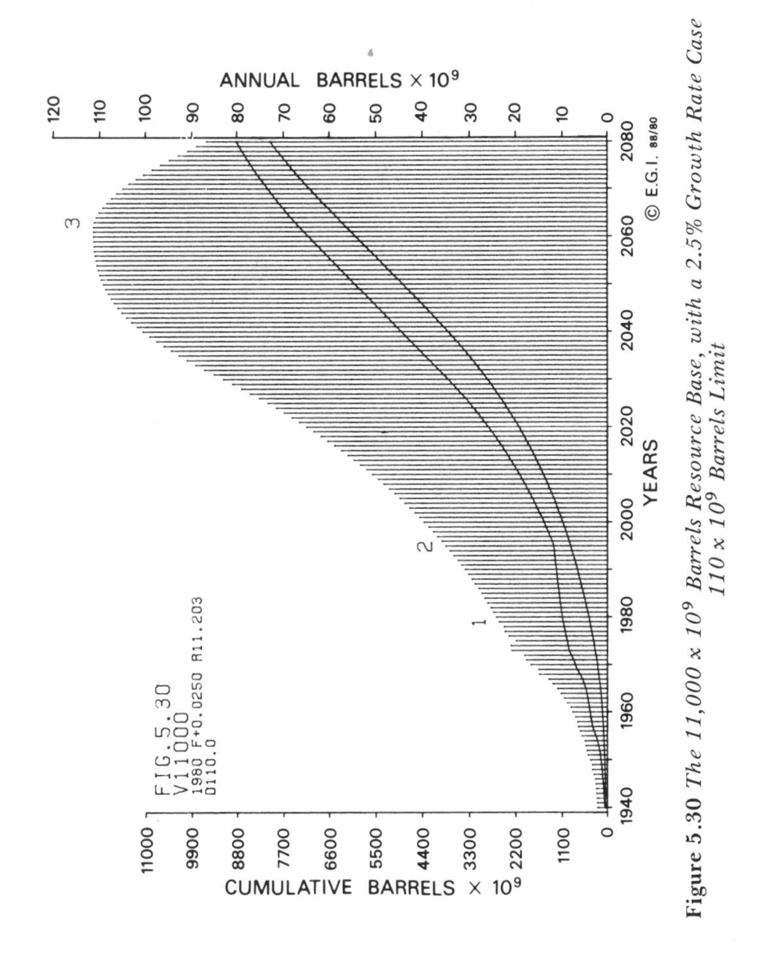

Figure 5.30 *The 11,000 x 10^9 Barrels Resource Base, with a 2.5% Growth Rate Case 110 x 10^9 Barrels Limit*

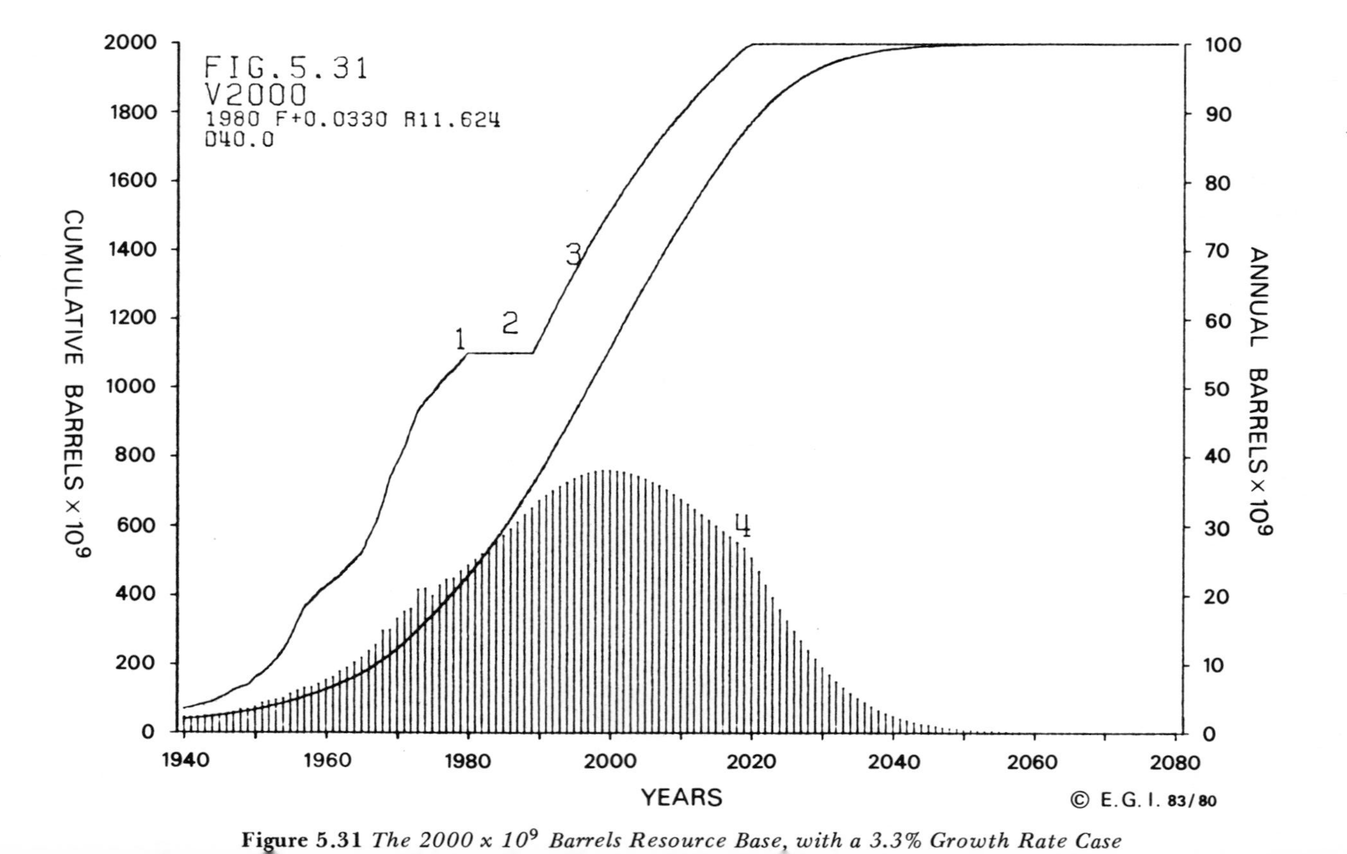

Figure 5.31 *The 2000 x 10^9 Barrels Resource Base, with a 3.3% Growth Rate Case*

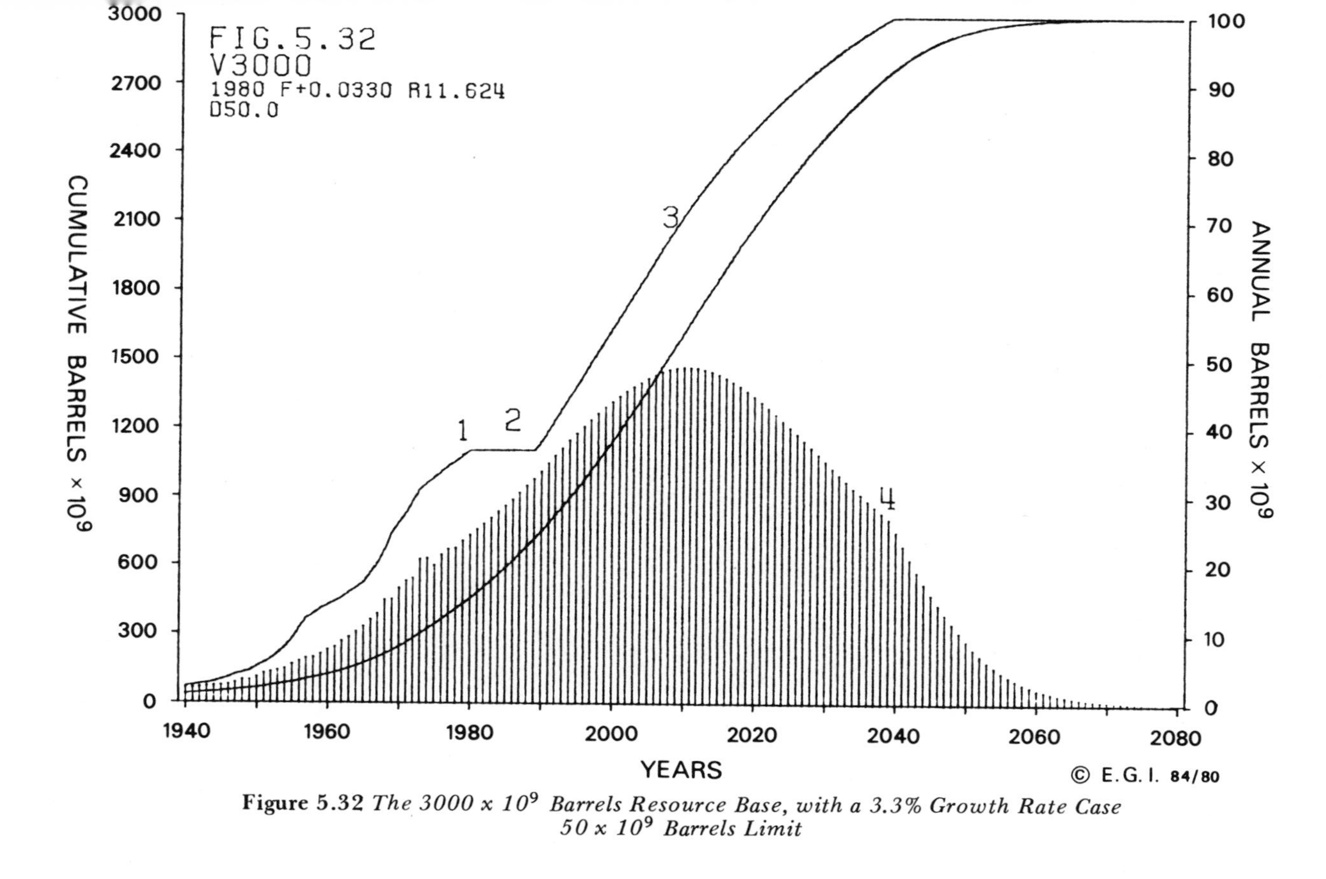

Figure 5.32 *The 3000 x 10^9 Barrels Resource Base, with a 3.3% Growth Rate Case 50 x 10^9 Barrels Limit*

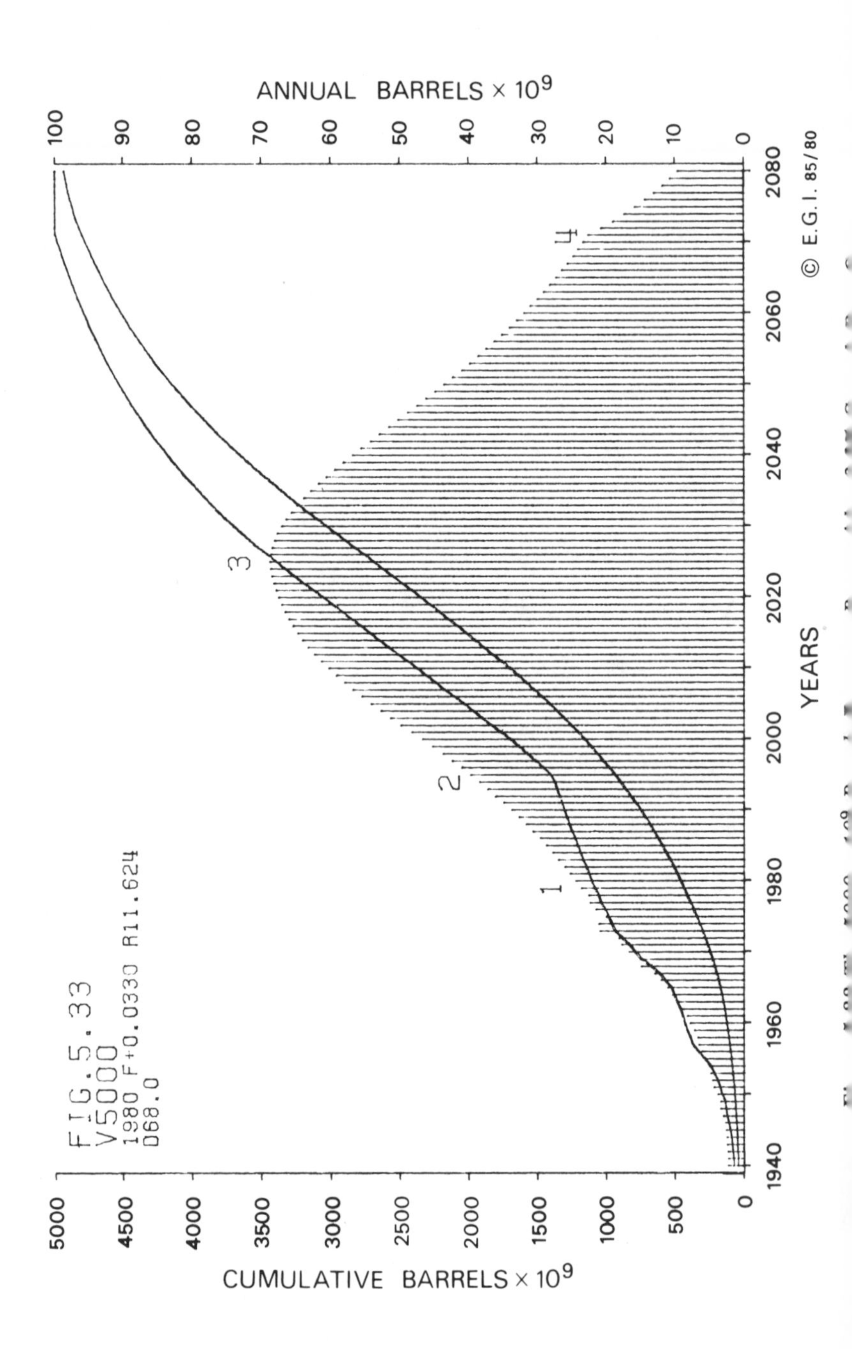
ANNUAL BARRELS × 10^9
100
90
80
70
60
50
40
30
20
10
0
FIG.5.33
V5000
1980 F+0.0330 R11.624
D68.0
1
2
3
4
5000
4500
4000
3500
3000
2500
2000
1500
1000
500
0
CUMULATIVE BARRELS × 10^9
1940
1960
1980
2000
2020
2040
2060
2080
YEARS
© E.G.I. 85/80

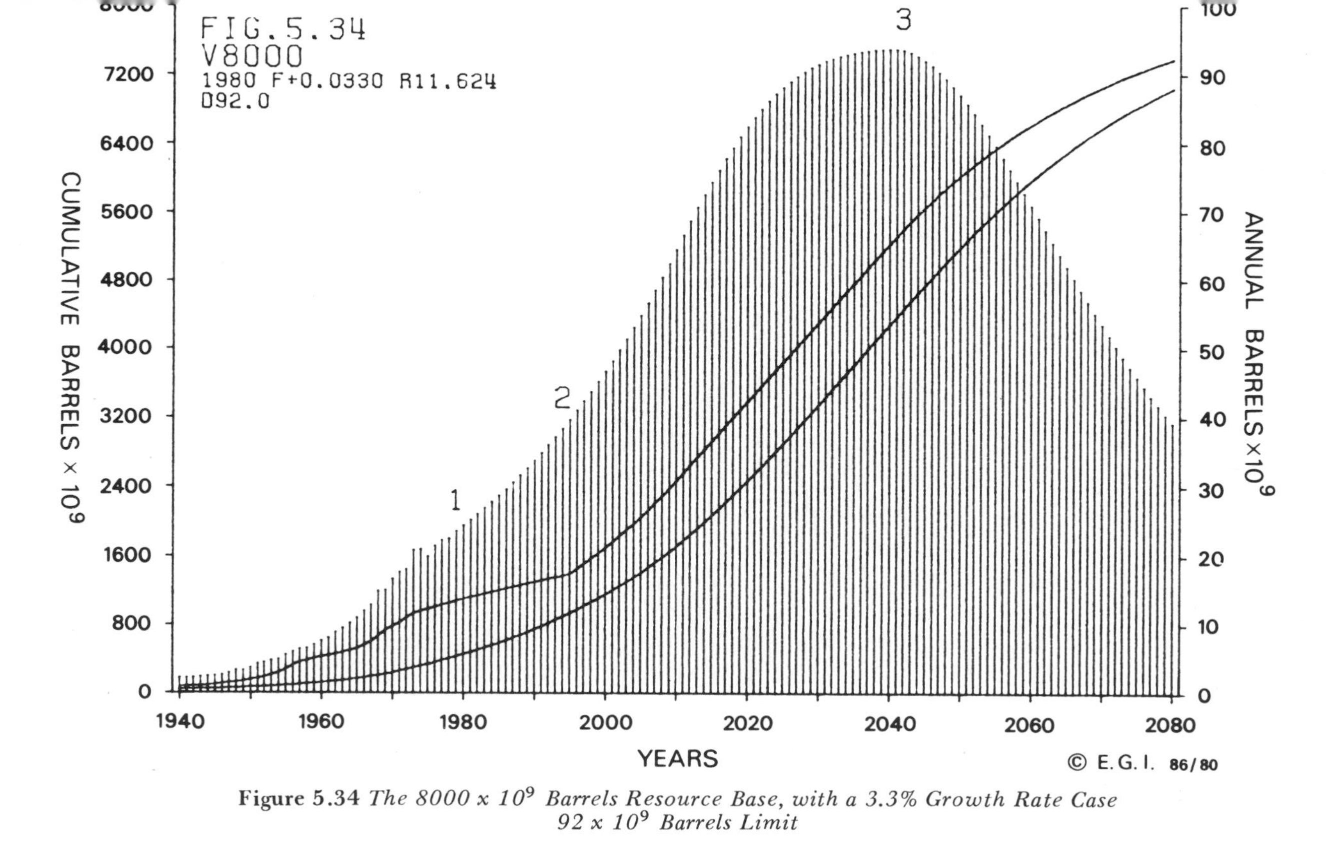

Figure 5.34 *The 8000 x 10^9 Barrels Resource Base, with a 3.3% Growth Rate Case 92 x 10^9 Barrels Limit*

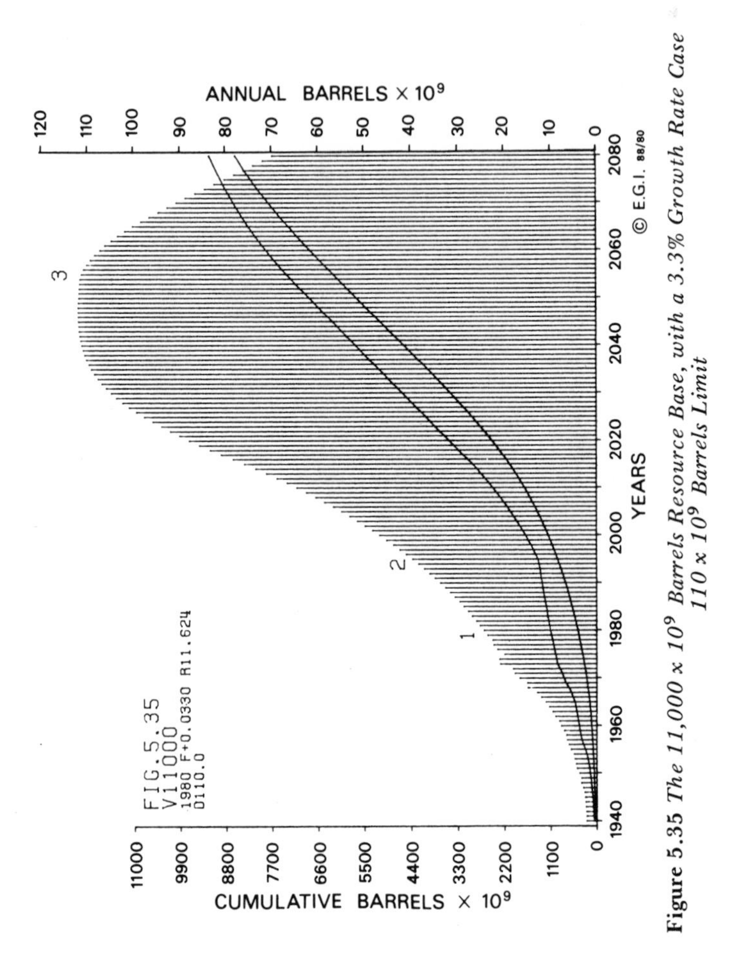

Figure 5.35 *The 11,000 x 10^9 Barrels Resource Base, with a 3.3% Growth Rate Case 110 x 10^9 Barrels Limit*

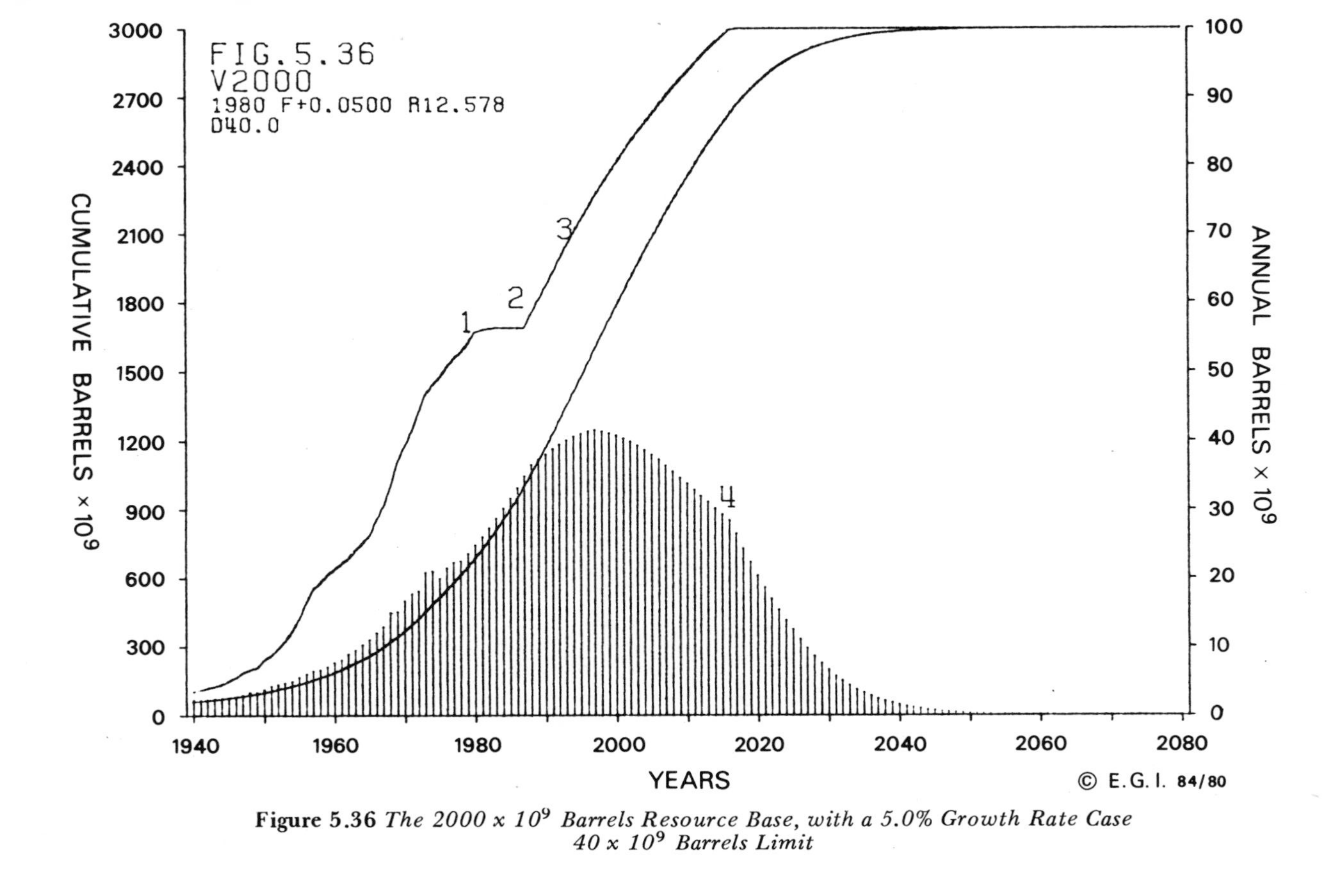

Figure 5.36 *The 2000 x 10^9 Barrels Resource Base, with a 5.0% Growth Rate Case 40 x 10^9 Barrels Limit*

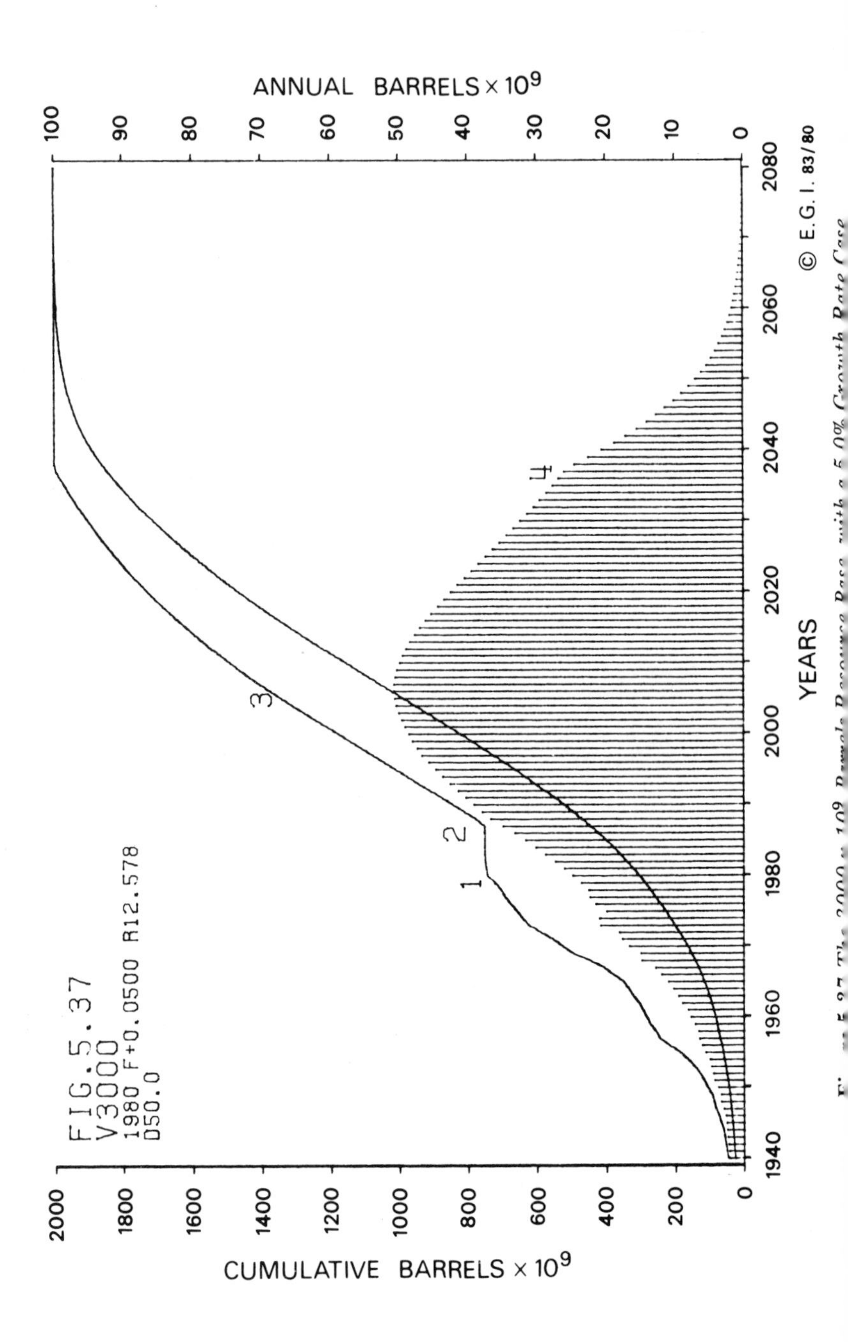

Figure 5.37 The 2000 × 10^9 Barrels Resource Base with a 5.0% Growth Rate Case

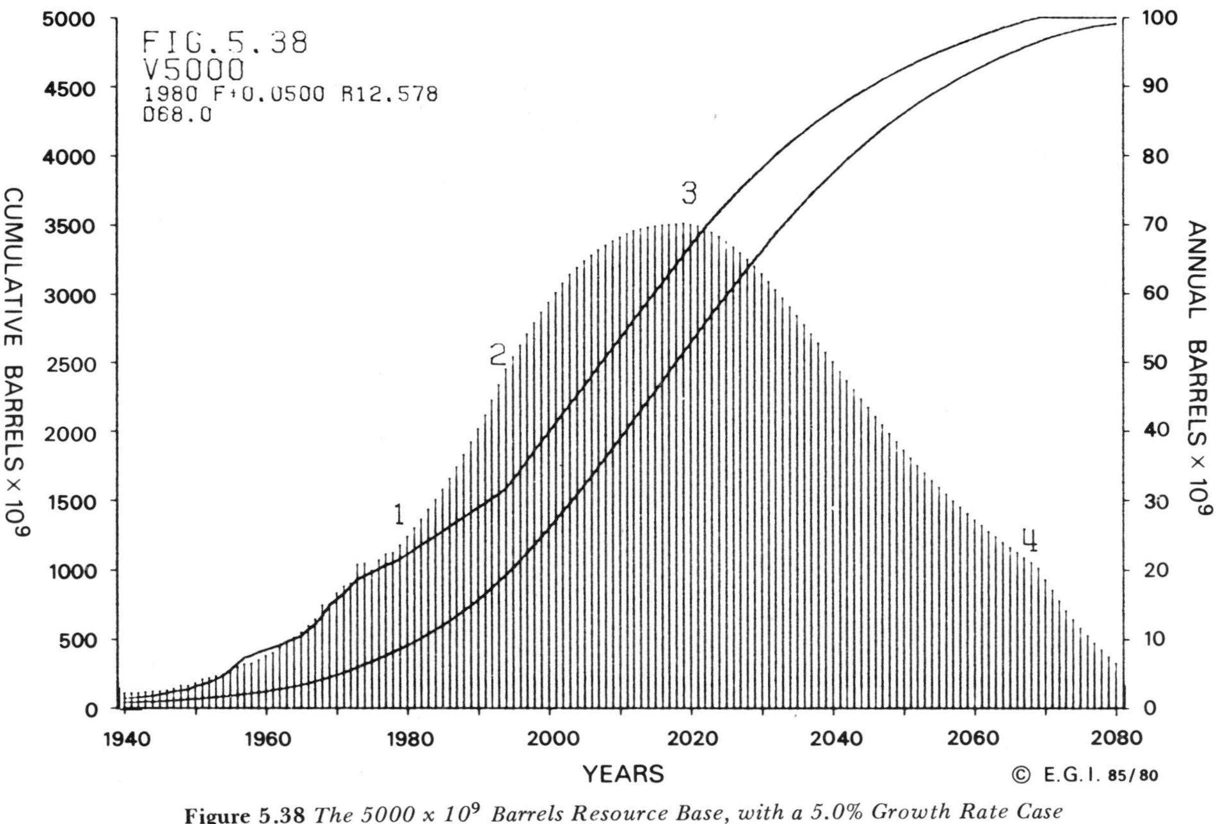

Figure 5.38 *The 5000 x 10^9 Barrels Resource Base, with a 5.0% Growth Rate Case 68 x 10^9 Barrels Limit*

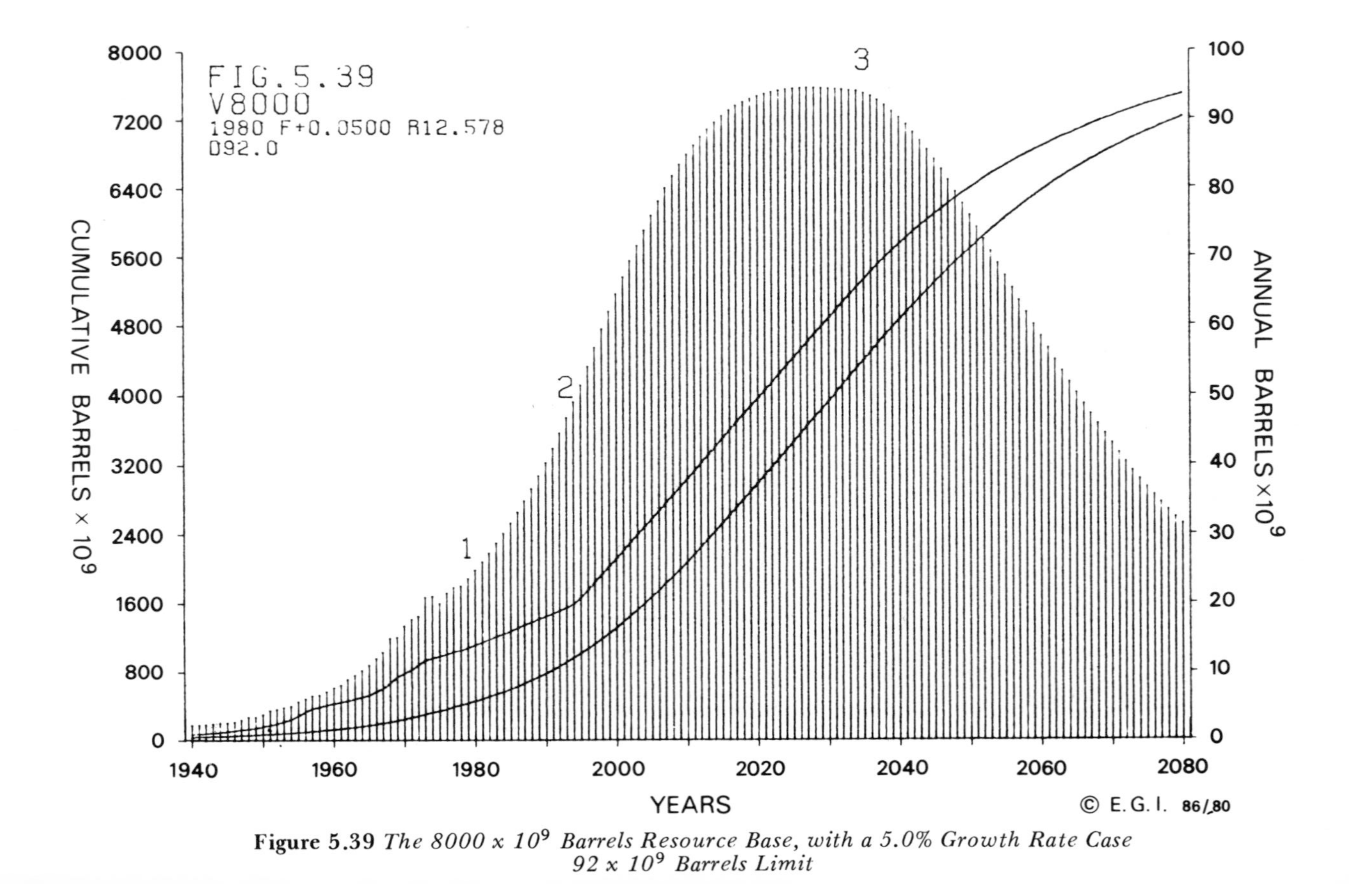

Figure 5.39 *The 8000 x 10^9 Barrels Resource Base, with a 5.0% Growth Rate Case*
92 x 10^9 Barrels Limit

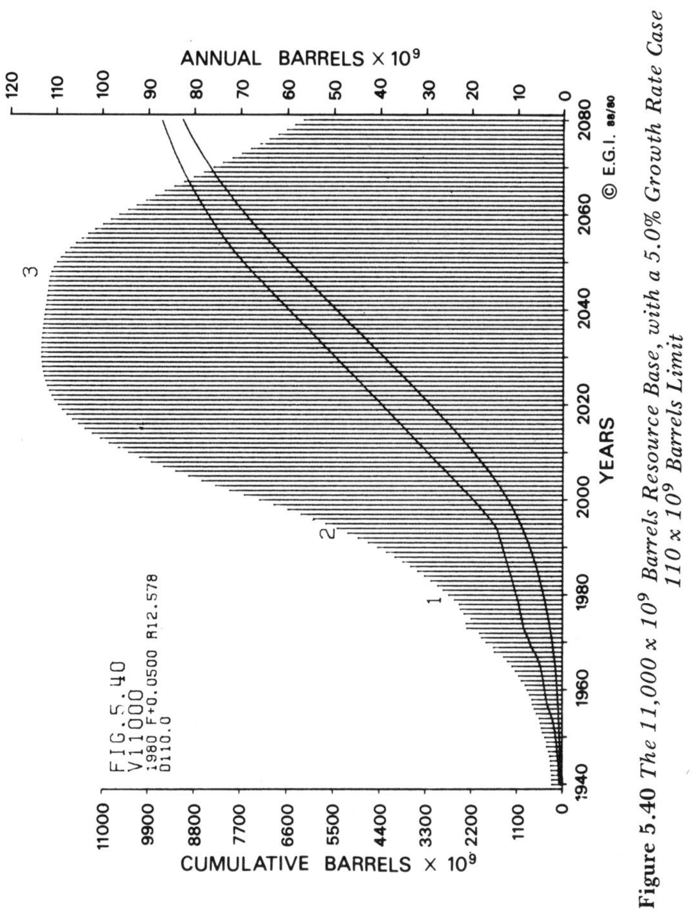

Figure 5.40 *The 11,000 x 10^9 Barrels Resource Base, with a 5.0% Growth Rate Case 110 x 10^9 Barrels Limit*

Chapter 6

Conclusion: International Policy Implications

A. Introduction

Our analyses and simulations of the future of oil have indicated that there is a 90 per cent chance that the industry could continue to grow until the year 2011. There is a 50 per cent chance that growth in oil production could continue until 2033 and even a one in 10 possibility that the industry would not reach its peak until 2072. Over the remainder of the industry's growth period the world-wide use of oil could increase to at least double and up to 3.3 times its present size (23.6 x 10^9 barrels annual production). Even after reaching its peak there would be several decades before the industry returned to its present level of development: namely, to 2039, 2077 and to well after 2080 in terms of the 90 per cent, 50 per cent and 10 per cent probability levels, respectively.

B. The Development of a Supply Crisis Syndrome

These conclusions are markedly at variance with the received wisdom of the moment as far as the future of oil is concerned. According to such wisdom the world faces an inevitable, near-future scarcity of oil. It must also be stressed, however, that the conclusions are equally at variance with the conventional wisdom on the future of oil of up to less than a decade ago. Prior to the last few years there was a general consensus on the ability of the oil industry to continue expansion at the exponential rate achieved over the period since 1945. This was then seen as likely to last so far into the future that there was no need for conservation or alternative energy planning.

In Chapter 3 we attempted to demonstrate the inherent irrationality of such supremely optimistic views on the future of oil, even in the context of what we have been convinced it is wise to consider as the largest ultimate oil resource base. It was, nevertheless, on the basis of that type of appraisal of a guaranteed future for oil in the long term that the world generally turned to an increasing dependence on oil. In this period of expanding oil use, the development possibilities of other energy sources were undermined and reasonable attitudes concerning the manner of oil use were overwhelmed by the belief in the desirability and inevitability of a 7½ per cent exponential growth per annum. In other words, policy decisions for the short term, from the mid-1950s to the early 1970s were based on an inappropriate interpretation of the longer-term prospects for the availability of oil. It is, in essence, from that earlier failure to appreciate the complex and dynamic inter-relationships between oil resources, reserves and use, that the world's current energy problems – the so-called energy crisis – really stem. In the first place we were completely disarmed in respect of the immense dangers inherent in a 10-year doubling rate in the demand for oil. These arose, not only because of the impact of such a development on the rate at which new resources of oil had to be found, but also because of the potential power that it gave to the oil producing countries.[1] Second, it produced a powerful disincentive for countries and consumers to use even known energy resources which were adjudged to be only slightly higher cost than the low – and declining – cost oil of the prolific oil-bearing regions, discovered and exploited in the 1950s and the 1960s. There was certainly no incentive to look for new energy resources – including undiscovered oil resources in non-producing countries – in most other parts of the world.

Thus, the world's increasingly oil-intensive economies and societies became dependent on oil reserves which were unduly concentrated in a relatively small number of countries. This was most marked, of course, in respect of the Middle East where, for reasons which originate essentially from the political relationships of the world's Great Powers in the 19th and early part of the 20th century, the oil industry had been strongly motivated to concentrate its attention. This region of the world appeared unique both in its geological characteristics as

1. One of the authors made this point – and attempted to quantify the potential costs involved, in an article published in February 1971. See P.R.Odell, 'Against an Oil Cartel', *New Society*, No.437, February 11, 1971.

a habitat for oil, and in politico-economic terms. It achieved the status of the happy hunting ground for the Anglo-American oil companies which, until very recently, were able to secure conditions there for the exploitation of oil which were generally unattainable throughout most of the rest of the world.

It is thus the 'accidents' of recent political and economic history which have come to determine the geographical pattern of the non-communist world's reserves of oil. This has had a profound effect on the present outlook on oil in that such a concentrated pattern of proven reserves has produced the belief that this pattern necessarily represents the ultimate geography of oil resources. Given contemporary conditions in the Middle East there is uncertainty concerning the future availability of the oil that has been discovered and exploited. Thus the oil scarcity syndrome of the last few years has evolved.

The international oil companies failed earlier to secure their acceptability in many parts of the world as a result of such factors as growing nationalism and the view of the oil companies as the main agents of economic imperialism. The companies have been upstaged since 1973 by the Oil Producing and Exporting Countries (OPEC), and have lost their control over most of the world's proven oil reserves as well as over the potential oil resources in countries whose oil industries they had created. This change in ownership and control over a commodity, essential to the West's modern economic and social systems, has produced the crisis of confidence over the future of oil.

Unhappily, it is in this atmosphere of crisis, arising out of the failure of the international oil companies, in their role as economic and political institutions, to maintain their position *vis à vis* the major oil-exporting countries, that policy decisions are now being generated. In essence, these have the effect of exacerbating the crisis. Indeed, they serve to turn it from a short-term crisis, which simply necessitated a limited period of readjustment by both the suppliers and the users of oil, to one in which the future of oil has been heavily discounted. This in spite of the fact that it is the energy source most appropriate for continuing to make both economic and social developments possible.

In other words, what were essentially short-term difficulties arising out of geo-political change, have become hopelessly interwoven with more important longer-term considerations. The present shortage of oil which really emerges out of the shift in the balance of oil power (from the international companies to oil-exporting countries) is now interpreted as a necessary long-term condition. As such it is producing policy decisions and actions which themselves serve to ensure that the very scarcity which is feared is, indeed, made a reality. As we have seen from the results of this simulation study, such a condition is unnecessary for a period of *at least* 30 years whilst, at the other extreme, it may not be a necessary development, from the point of view of the dynamics of resources, reserves and demand inter-relationships, for another 100 years.

C. The Cost of Oil

Acceptance of the hypothesis that a near-future physical scarcity of oil is not a necessary condition on which to determine the evolution of the world's economic, social and political order, opens up a range of possibilities which are otherwise excluded. From the point of view of the cost of energy it casts doubts on the belief that there is only one way the price can go, namely 'up', even from the dizzy heights that have already been secured by the countries currently controlling the supply, and hence the price. Hard though adjustment to $35 per barrel oil may be amongst consuming nations and consumers there is currently a feeling of inevitability about it. This emerges from the view that the rapid depletion of the world's remaining limited oil resources justify such a price. The same attitudes imply that we must be prepared to face a relatively near-future of even more expensive alternative energies – notably nuclear power, the so-called attractiveness of which is basically a function of the belief in the scarcity of oil.

There are even serious doubts expressed over the validity of what appears to be a highly reasonable proposition on the evolution of alternative energy costs, *viz* that it is quite reasonable, in the light of historical experience, to hypothesize a fall over time in the real costs of alternatives to conventional oil. Where such alternatives are already in development, progress can be made along the learning curve towards lower technological costs, and economies of scale can be increasingly realized as the opportunity to move to bigger units is generated. Fisher[1] has shown that these factors have always worked in respect of

1. John C. Fisher, *Energy Crises in Perspective*, J.Wiley and Sons Inc., New York, 1974.

the historical processes of energy resources development. Indeed, the essential requirement for the success of the prospects for the development of alternatives to oil, is time. Time is required for the learning process, as well as for the achievement of technological break-throughs on the basis of which reduced costs can be achieved. The real, as opposed to the feared, prospects for oil not only give this necessary breathing space, but also provide the opportunity for a smooth transition to alternative energies at unit prices which, in real terms, could fall to well below today's $35 per barrel oil.

D. Disincentives to Oil Supply

In the meantime, except in the context of a world supply of the commodity which is controlled by a limited number of countries with production to spare for export, $35 per barrel is well above the long-term supply price for conventional oil production. This is true almost irrespective of its habitat (including deep-water offshore), of its reservoir characteristics (in terms of size and depth), and of difficulties in recovering it as a result of specific physical and chemical variables in its occurrence. Revenues of $35 per barrel, available in total to the potential producing company, would give a return on investment more than adequate to justify a 'go' decision in almost every case.

However, in the context of the general belief in oil as a scarce commodity and given the ability of governments to lay effective claim to all or most of the economic rent emerging from the production of relatively low-cost oil and its sale at high prices, the producing companies cannot lay their hands on anything like the revenues generated by $35 per barrel oil. The effectiveness of governments' claims to the economic rent emerges from the idea that the depletion of a country's 'scarce' resources can be justified only in terms of the revenues they earn for governments. On the basis of such government revenues the 'oil can be sowed' (to use the Venezuelan term) to diversify and strengthen the economy. The consequence of such attitudes and actions is to eliminate a greater part of the motivation for companies to seek more oil, and/or to produce more of that which they have found. This is another element in the man-made process of creating a scarcity of oil.

In such circumstances it is hardly surprising that the oil

companies prefer to diversify into other sorts of energy which, though inherently more expensive to produce, usually give the prospects of a better return on investment. Governments, to date at least, have not had the same propensity to intervene in their exploitation and development. Even worse, from the point of view of the world's energy outlook, the companies prefer to diversify their investments away from energy altogether and into activities in which governments seem even less likely to get involved. This, of course, works to the detriment of the potential supply of the energy that may be required to complement the now slowly increasing need for oil.

Thus, unhappily, not even $35 a barrel oil in 1980 necessarily enhances the supply potential. As a consequence of the behaviour of most governments – both in terms of their belief in oil as an inherently scarce commodity and in the light of their attitude that the production of oil can be used to ensure a flow of substantial government revenues – the supply of oil seems likely to remain much less elastic than might reasonably be concluded, were this to be viewed simply as a proposition in economics.

This is not the only, or even the most important, way in which government intervention inhibits the development of a supply of oil in the quantity which could otherwise be expected with oil at its present price. There is also the question of governments' price expectations and the low discount rates they tend to use in order to compare present with future alternatives. If oil is scarce (again the first and basic assumption of the analysis in which energy planners indulge), so that the price will rise over time, it is easier to demonstrate that it is preferable to keep the oil in the ground, rather than to produce it. Hence, the search for, and the hope of the achievement of, optimum national oil depletion rates by governments as dissimilar as those of Kuwait, Britain, Canada, Norway and many others in different parts of the world. The end results of this factor are: first, less production in the immediate future (hence exacerbated short-term scarcity and consequential politico-economic problems of the kind which now so beset the Western economic system); second, less development of known oil provinces (and hence the propensity to create a medium-term scarcity of oil); and, third, less exploration (with its inevitable consequences *viz* the achievement of a long-term scarcity of the commodity, simply because potential supplies have not been intensively enough sought).[1]

There is yet another set of forces at work, in the context of national energy planning activities, which serves to help make the belief in scarcity more of a reality. This is the set of forces concerned with the feared adverse effects of oil exploration and exploitation on the environments of particular countries. It includes concern for pollution – both of water and atmosphere; for the social effects of oil developments on specific communities and on particular localities; for the impact of oil developments on other economic activities which may have to compete with the oil industry for scarce resources such as labour; and, finally, concern for the macro-economic effects of oil export earnings and of oil revenues on the structure of a country's economy. These are all basically management and control problems and none are incapable of solution in sophisticated Western industrialized societies, even with rapid and extensive oil exploration and exploitation activities. They are, nevertheless, problems which can be, and, indeed, sometimes are used as reasons for restricting the rate at which oil developments are permitted, especially in Western Europe. Unhappily, in the enthusiasm for the short-term justification for such restrictions, frequently generated by particular vested interests, the longer-term issue of their cumulative effect on the outlook for oil is usually ignored. If treated at all, it is relegated to a minor role in the decisions. The consequence is a contribution to the creation of oil scarcity in the medium- to the longer-term.

E. The Impact of the Structure of the Industry

It is not only the governments of oil producing countries (or of potential producing countries) which contribute to the likely achievement of a scarcity of oil in the long-, as well as the short-term, and to the creation of a higher-than-necessary price for oil as a problem of the moment. The future development of

1. This multi-faceted oil play effectively describes the state of the game in Western Europe where known reserves are deliberately not produced; where oil and gas provinces are deliberately restricted in terms of their development; and where exploration is deliberately discouraged. And yet the governments of Western European countries express their deep concern over the prospects for future supplies of oil. For a further discussion of this paradoxical situation see, P.R.Odell, 'Towards a More Rational View of the Energy Policy Options open to the Western Oil Consuming Nations', in: *Energy – What Now?*, K.E.Davis and P.deWit (Eds.), Bonaventura, Amsterdam, 1979.

the supply and price of oil is also a function of the structure, the organization and the financing of the industry. As mentioned earlier in the chapter the international oil companies, which have hitherto been responsible for most of the technological progress and investment in oil, have become decreasingly acceptable in many parts of the world as instruments for the exploitation of oil. In the short- to medium-term this inevitably limits the degree to which new oil can be found and produced and the extent to which already discovered and producing fields can be further developed with enhanced recovery facilities. This is a function of the fact that expertise remains concentrated in the hands of these companies and that, to date, there has been inadequate development of any large-scale and effective alternative to the companies for the exploitation of most of the world's oil resources. In addition, even in areas in which the companies continue to work they do not necessarily find it commercially worthwhile to produce oil as intensively or as extensively as the resource base allows. In the newly developed North Sea oil province, for example, many fields with less than 200 million barrels of estimated recoverable reserves, together with parts of many larger fields, from which oil cannot be recovered through a production system which gives an adequate return on investment, have been left unexploited. This is because the companies responsible for their discovery/ development have determined that the investment necessary to produce their oil would not generate a rate of return high enough to justify its being made.[1]

In such areas and conditions, and particularly in large parts of the Third World where, as Figures 1.6 and 1.7 show, much of the world's remaining oil potential lies, the future development of the world's oil resources depend, to a considerable extent, upon a changed structure for the organization of the oil industry. Without a change, there appears to be little chance of ensuring a large enough flow of expertise and investment into oil exploration and development so that the world's oil resource base – whether of 3000 or 11,000 x 10^9 barrels – could be adequately exploited over the next 30 to 100 years.

There is thus a further basic paradox in the world of future oil potential. The oil companies, imbued with a sense of political realism arising out of their long experience, now

1. See P.R.Odell and K.E.Rosing, *The Optimal Development of the North Sea Oilfields*, Kogan Page, London 1976, for an extended discussion of these issues.

discount their opportunities to exploit much of the remaining potential. In their public presentations of this 'fact of life', they help to create and sustain the idea of oil as a scarce commodity. This is seen, for example, in their views on the potential resources of Latin America, Africa and South East Asia. Earlier in the book these were contrasted with the views of other organizations (see Table 1.1). But the companies are viewed generally as the best source of information (even as objective sources of information) on the prospects for oil, in spite of the hostility often shown towards them. Therefore their commercially inspired interpretations are accepted and, in being accepted, enhance the inherent scarcity belief.

In other words, in spite of their undoubted expertise and increasing technological capabilities, both in finding and producing oil, the companies are responsible for much of the pessimism and the all-pervading lack of confidence in respect of the future of oil. Yet, as we have shown in Chapter 5, and as the companies themselves at one time stressed (see Chapter 3), such pessimism and lack of confidence is by no means a necessary part of the outlook for oil; and as oil remains the main source of energy for a developing world this is an important issue. The massive swing in the companies' attitudes to the future of oil – from the supreme optimism and confidence of the period up to the oil crisis (see Chapter 3) to their present pessimism and lack of confidence (see Chapter 4) – presents a sobering contrast.

F. The Future of Oil

The evolution of a future of oil which could relate closely to the statistical probabilities based on the inter-relationships of resource base size, adequate additions to discovered reserves year after year, and to a modestly increasing, and more efficient, use of oil in the development process,[1] depends upon three inter-related factors. First, the elimination of the present crisis mentality in respect of the longer-term availability of oil. Second, the willingness of governments, particularly those of the OPEC and of the OECD groups of countries, to sustain and

1. We shall not discuss the question of the efficient use of oil any further in this study. As we have shown, the rate of increase in use is an important component in looking at the future of oil. Other studies show the immense scope that exists for so enhancing the efficiency of oil use that the total used need not increase above today's levels even in a growth economy. See, for example, G.Leach et.al., *A Low Energy Strategy for the United Kingdom*, IIED, London, 1978.

encourage the exploration and exploitation of their countries' oil resources. And third, the establishment of an adequate organizational infrastructure to make the required exploration and exploitation efforts possible in all parts of the petroliferous and potentially petroliferous world.

Important though the first two of these factors are for the future of oil, especially in the short- to medium-term, it is the third factor which appears to constitute the critical variable for the longer-term outlook. Any success achieved in the establishment of an infrastructure appropriate for the future exploitation of all the world's oil resources would automatically generate conditions in which there would be changes in attitudes to the availability of oil and in government policies towards its development.

The under-statement of the world's oil potential is, as we have stressed, currently justifiable from the point of view of the international oil companies. Such commercial justification for contemporary under-statement arises in exactly the same way as that of the companies' over-statement when their commercial viability depended on their success in selling rapidly increasing quantities of oil in the context of energy markets in which there was plenty of scope for oil to substitute other sorts of energy (for example, coal in Western Europe and the United States and vegetable fuels in Latin America – see Tables 1.3 and 1.4, respectively). In today's changed international economic and energy market conditions, however, the effective resource base for the companies is a more important consideration. And for them the effective base is the one they expect to be able to exploit. If their relations with many countries of the world are either bad or non-existent and the companies cannot conceive of circumstances in which large investments for oil development could be committed to them, then they have no option but to discount the resources concerned. The companies then visualize the world's energy future as one in which the output of oil will soon reach its peak and thus need to be supplemented by the use of coal and nuclear power etc. This is the kind of view which Shell and BP have recently tried to put across in their public relations' type of advertising (see Figures 4.1 and 4.2).

The fact that the Western world currently depends on institutions which, by virtue of their very nature, cannot find and develop much of the world's oil, does not mean that the resource base cannot be developed. The international oil companies are a set of institutions capable of being changed, and/or substituted and/or complemented. New or revised institutional arrangements for tackling the opportunities provided by the under-explored and little developed parts of the world can be created, within a decade or so. Such a development would be in time to take up exploration for, and exploitation of, the world's oil resources prior to the reserves/production ratio falling to so low a level that confidence in the future of oil is completely undermined. Our 50 per cent probable future of oil (see Figure 5.46 and pp. 192 to 196) showed that it will be 1995 before the R/P ratio falls to the level at which the industry has to find more reserves. There is thus breathing space available for action to be taken, even beyond the minimum likely period necessary for the creation of a revised institutional framework, capable of ensuring the long-term development of the industry.

G. A New Role for the Companies

Given that successful oil development depends on the continuing availability of managerial and technological expertise and on large-scale financial resources; and that 90 per cent – or possibly more – of such resources exist within, or are only available to, the international oil companies and a limited number of other entities, especially in the United States; then an expanded involvement of these companies in the future of oil would appear to be a prerequisite for success. For these companies to give less attention, or to be forced to give up, oil exploration and production, could well be the means whereby an otherwise achievable long-term future of oil is undermined.

Thus, the future of oil may well depend on finding a way in which the international oil companies can become associated with exploration and production in as many countries of the world as possible. This, of course, means that their involvement has to be politically acceptable to the countries concerned, whilst from the companies' point of view it has to be commercially attractive.

Somewhat paradoxically, in the light of the recent nationalization of their oil industries, it is in the traditional oil producing and exporting countries that the need to re-involve the oil companies has already been recognized and acted upon. This has been done as a means whereby the countries can secure access to the companies' expertise, even if not to their financial resources as these hardly constitute a problem for the member countries of OPEC, with oil at $35 per barrel. In many OPEC countries agreements have been signed with the oil companies whereby the latter, in return for a fee (usually related to production levels) offer managerial and technological knowledge and services to the state oil company, so that it will be in a better position to ensure the continued discovery of new fields and of new oil producing regions for the longer-term future.

Such arrangements avoid the earlier political unacceptability of the companies and provide a way whereby their accumulated expertise can be tapped. The fact that major oil producing and exporting countries can negotiate this sort of agreement is an indication of their new-found politico-economic strength and of their confidence in their ability to control the situation in their own interests. The companies, for their part, achieve contracts which are virtually without risk, in that payments are related to existing levels of production and are made by countries in which government revenues are plentiful.

It is difficult to imagine that this sort of country/oil company relationship could develop quickly in respect of the opportunities for oil exploration and development in the world's less developed countries. There are several reasons for this. First, the necessary parity of esteem does not exist between the two parties to enable effective managerial/technological relationships to be established. Second, most less developed countries lack the necessary knowledge of the oil industry to handle the complex negotiations with confidence. Third, the oil companies have insufficient confidence in the opportunities offered by such countries, most of which lack the necessary infrastructure to make oil activities easy to develop. Fourth, and most important, the countries concerned usually lack the necessary and very considerable financial resources to invest in the high risk business of finding and developing oil. The oil companies are aware of this and rate their chances of being paid for efforts to find oil, which turned out to be unsuccessful, as rather low.

H. Investment in the Future of Oil

A recent World Bank report[1] has very conservatively indicated

1. World Bank Report No.1588, *Minerals and Energy in the Developing Countries*, Washington, May 1977.

an immediate potential availability of at least 60 x 10^9 barrels of oil in those parts of the Third World in which the international companies, for the reasons set out above, find it difficult or impossible to work. The report indicated the opportunity to develop a production of over five million barrels per day within a 10-year period, given a level of investment in exploration, production and associated transport facilities which, in 1976 dollar terms, would require upwards of $6000 million per year in investment. Needless to say, in relation to this size of annual financial requirement, the World Bank also indicated that it could not help, except in a very minor way. It would not be able to lend money for exploration because it was too risky, and even in terms of investment for field development and transport facilities the World Bank would be limited to undertaking two projects per year, each involving only $300 million. This, in essence, is an indication of a near zero ability of the Western world's main international financing agency to help develop the Third World's oil resources.

There is thus a formidable inability on the part of exisiting private and public international institutions to do anything significant to ensure the development of most of the world's remaining oil resources, in spite of the fact that the commitment of funds and of other resources to such development seems more likely to produce more energy per dollar invested than the investments made in the search for alternatives. This is true whether this search involves renewed interest in old oil regions, in order to try to recover a little more of the oil in place, or investment in other sources of energy. Nevertheless such alternative investment strategies are being followed. Their imposition is a result of the inadequate contemporary institutional facilities for finding and developing lower cost oil reserves and thus represents a grossly sub-optimal allocation of the world's limited financial and technical resources in the search for energy.

This development is not in the long-term interests of any of the parties involved. Both the industrialized (the main oil consuming) countries and the present oil producing (and exporting) countries, as well as the poor Third World countries, have an interest in the creation of a system in which the world's oil resources can be sought and developed at a rate which is high enough to sustain the growth in use needed to keep the world's economy moving ahead. The responsibility for ensuring this is shared by the two first named groups of countries, whose joint ability to provide a level of funds capable of sustaining the required effort is a matter of political will, rather than of economic ability. Given their co-operation, perhaps under the aegis of a joint OECD/OPEC agency, the international oil companies and other appropriate institutions could be hired to manage and/or to undertake oil exploration and development work in Third World and other countries.[1] As a result, the companies would, on the one hand, be 'de-politicized'. They would be hired simply for a specific exploration and development effort, in the context of which they could not exercise any political or economic power and influence. On the other hand, the companies themselves would be confident of being paid for their services and so be able to judge the requests made to them in a normal commercial way. The end result would be a continuing and a geographically much more dispersed pattern of oil exploration and development. This, on being implemented over the long term, could ensure a future of oil more closely approximating the opportunities offered by the world's ultimate resource base. It would provide a 90 per cent probability that oil will remain a growth industry into the second decade of the 21st century, and a 50 per cent probability that oil will be able to meet much of the world's modestly growing needs for energy until almost the middle of the next century.

1. The establishment of such an Agency, in the context of the new international oil situation and the potential for its development, has been fully discussed and evaluated in P.R.Odell and L.Vallenilla, *The Pressures of Oil*, Harper and Row, London, 1978.

Index

Index